NIGHT FIGHTER'S HANDBOOK

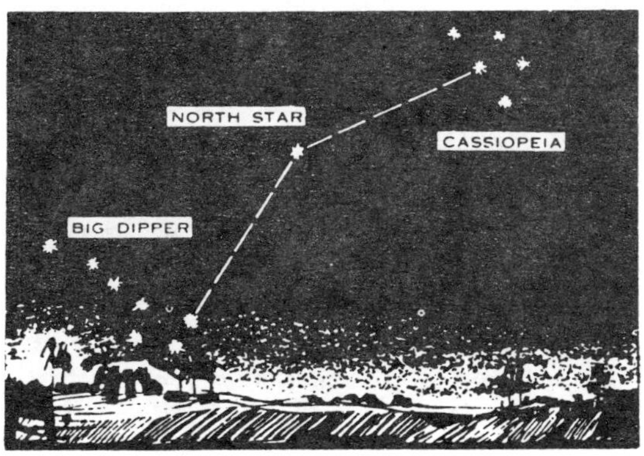

MAJOR DENNIS J. POPP

PALADIN PRESS
BOULDER, COLORADO

Night Fighter's Handbook
by Dennis J. Popp

ISBN 0-87364-361-5
Printed in the United States of America

Published by Paladin Press, a division of
Paladin Enterprises, Inc., P.O. Box 1307,
Boulder, Colorado 80306, USA.
(303) 443-7250

Direct inquiries and/or orders to the above address.

Contents

This book is dedicated to the officers and men of the U.S. Army's new elite, the "light fighters" of the 7th Infantry Division (Light) and their commander, MG William H. Harrison.

Foreword

In order for the light infantryman to effectively fight heavier opponents, he must make maximum use of all possible advantages. One of these advantages is fighting at night.

Historically, throughout the world, the night fighter—the "master of darkness"—has been greatly feared. The night fighter's own effectiveness, coupled with his opponent's fear of the dark, has often resulted in spectacular success in battle.

However, the ability to fight well at night is a *learned* skill. This manual is published to help light fighters become *night fighters.* It is organized into eight chapters that cover general night operating considerations and specific techniques. The last chapter is devoted to training techniques to help the small unit leader train his personnel for night operations.

Throughout this book, examples are provided to help the small unit leader use innovation and initiative in solving problems. Whenever confronted with a new obstacle, the light fighter should immediately begin the problem-solving process. The problem-solving process consists of the following steps:

1. *Define the problem.* A good, definitive statement of the problem will suggest solutions by itself.

2. *List alternative courses of action.* List as many alternatives as possible that may solve the problem.

3. *Analyze the courses of action.* For each course of action listed, determine its positive and negative aspects.

4. *Select the best course of action.* The best course of action will solve the problem with the most positive and fewest negative effects. Practicing the problem-solving technique will allow you, the small unit leader, to become adept at finding creative solutions to problems you encounter in field training and combat environments.

HARRY W. MOHR
LTC, IN
Commanding

Remember: *While the sheep huddle together at night, the tigers stalk.*

Preface

Light fighters are the U.S. Army's answer to increased small-unit firepower without an extended logistics train. The U.S. Army has airborne infantry, mechanized infantry, and light infantry modeled after Ranger and elite units of World War II. Light infantry embodies high firepower in a small infantry unity with almost no logistics capability. A light infantry unit is a combat unit supported almost entirely by other organizations. Its job is simply to fight, while other organizations are concerned with food, ammunition, and transportation. The light infantry fights on the ground, takes the ground, and wins the battle. This manual is dedicated to light fighters and is designed for their use and training.

The information in this manual was extracted from unclassified documents found at Forts Benning, Hood, Leavenworth, Lewis, and Ord and from the author's personal papers. Most of the information was checked for practical value and/or supplemented in oral interviews with combat-experienced officers and NCOs at Forts Leavenworth and Ord.

The research and initial planning for the manual was completed during training tours for reservists in January and March of 1985; the manual was coordinated and drafted later.

This manual is intended to be informative only; its message should not be construed as U.S. Army doctrine. In short, the manual does not pretend to supply the "school solution" to night-fighting problems, only to provide a range of solutions that have been successful in the past. It includes lessons learned by four nations' land forces in combat during conflicts between and including WWI and Viet Nam. The solutions it provides were evaluated by infantry leaders in the grades of sergeant through lieutenant colonel.

Many, many soldiers and civilians contributed to this manual. I am particularly indebted to:

- COL John D. Howard, Cdr, 2d Bde; LTC Harry W. Mohr, Cdr, 1/32nd Inf; SSG Gordon and SGT Gilford of 1/32nd Inf(L) for their suggestions;

- MAJ Robert Clark, IRR, for advice and assistance in editing and rewriting;

- 1LT Chapman, SSG Spence, and SP4 Kibby for the monumental administrative effort, resulting in legible copies for coordination and review, while also functioning in a line–battalion PAC;

- Mrs. Crow, research librarian at Fort Leavenworth;

- and to the Night Fighting Project officers at Forts Hood, Leavenworth, Benning, Greeley, and Lewis who so freely shared their data.

List of Abbreviations

AO – area of operation
AW – automatic weapon or automatic weapon carrier
BDU – battle dress uniform
FDC – fire direction center
GSR – ground surveillance radar
IF – indirect fire
IR – infrared
MG – machine gun
METT-T – mission, enemy, terrain, troops, and time
NVD – night vision device
OCOKA – observation, cover, obstacles, key terrain, and avenues of approach
OP – observation post
ORP – objective rally point
RP – rally point
RTO – radio telephone operator
SOP – standard operating procedure
WP – white phosphorus

1. Characteristics of Night Operations

Generally, the physical and psychological effects of darkness upon the human body and changes to the environment are what give the night operation its special characteristics. Understanding these effects is the key to being able to master the darkness.

PSYCHOLOGICAL EFFECTS

Darkness increases fear and a sense of foreboding, leading to imagined dangers and panic under sudden stress. This effect is very well overcome by developing a sense of belonging to a team and of not being alone. Thus, being part of a buddy team and getting frequent visits by a seemingly confident squad leader will greatly decrease the negative psychological aspects of night operations on a light fighter.

Confusion and indecision by leaders about unrelated problems leave men far more prone to illusions and panic. For example, a squad leader worrying about home problems may unwittingly cause his men anxiety about night battle. Be deliberately confident!

The ability to function and fight at night is directly related to confidence in personal skills, confidence in the unit, and confidence in the leaders. Confidence is built through guided practice.

PHYSICAL EFFECTS

Seeing

At night you see with special cells in your eyes called rods; these cells don't differentiate color well and "burn out" easily when exposed to light. They need about thirty minutes of dark or dim red light to activate. There are fewer of these cells in the center of the eye (see Figure 1). Images that you concentrate on, that is, look directly at, tend to disappear because you are using the center of the eye with fewer night vision cells. Or, the image may appear to move. In extreme darkness you may see imaginary stationary or moving lights. (This hallucination confused the early astronauts; it appears to be due to "overcharging" of unused optic nerves, but is not harmful.) To see at night, relax and scan methodically, using off-center vision (see Figures 2 and 3).

Alcohol, nicotine, carbon monoxide, fatigue, and, possibly, a high-fat meal, degrade vision. Lack of oxygen also degrades vision; pilots cannot see well enough at over 4000 feet to consistently land safely. One cigarette produces the same effect as an increase of 3000 feet in altitude (because of less oxygen in the blood and because nicotine constricts the blood vessels, including those in the eyes, reducing circulation). Exces-

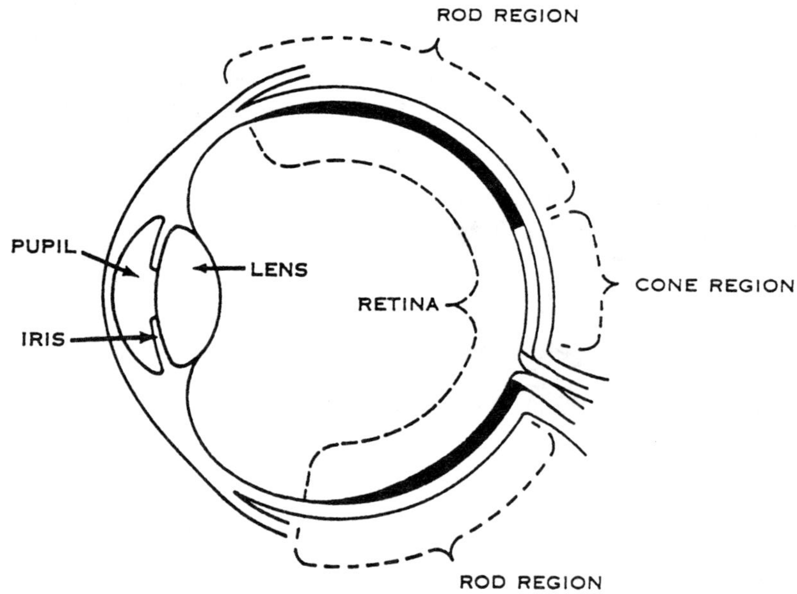

Figure 1. The eye.

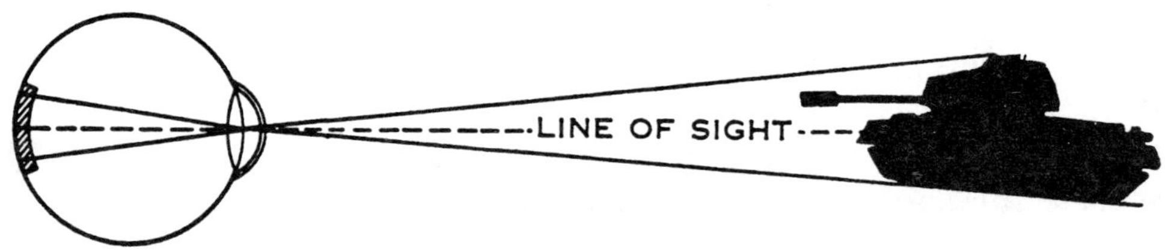

Figure 2. Day vision. Look directly at the object so that the image is formed on the cone region.

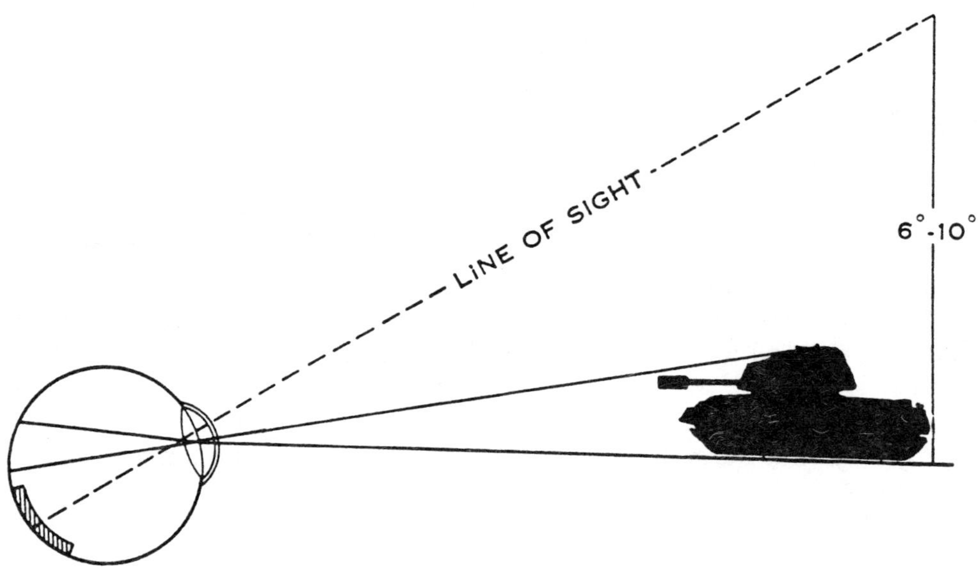

Figure 3. Night vision. Look slightly away from the object so that the image is formed on the rod region.

sive sunlight degrades night vision for about thirty-six hours; so wear sunglasses during the day if you are going to be operating at night, if possible. This is particularly necessary during desert operations.

Distance perception at night is distorted. Small objects seem farther away, large objects closer, since we judge distance partly by size relationships, and other objects may not be visible in the dark. Also, bright objects seem closer than dull objects (important info for night recons). And, forests may look like hills at a distance in dim light. In WWII, Japanese night tactics became somewhat less effective when allied forces used NVDs (night vision devices)—and much less effective when the Japanese used new or poorly trained troops. Training is the key to night effectiveness.

Binoculars can improve night vision by nearly as much as the older NVDs. A large front lens and exit pupil such as those found on U.S. Navy binoculars are most helpful for night vision assistance, but U.S. Army binoculars are also very effective. Also, using binoculars with PVS-5s, a night vision device, greatly extends your range of night vision.

Hearing

Hearing becomes more acute at night, probably because we concentrate more, but also because of a lower "background noise" level, and colder, moister air carries sound better.

The biggest problem with hearing at night is the soldier's lack of confidence in this sense. This can be overcome with a lot of directed practice. Soldiers can double their ability to discriminate multiple sounds, faint sounds, and sound-source direction with about fourteen hours of directed practice.

A rifle-bolt closing can be heard for a long distance. Have your men load their weapons early, wait until firing starts to clear a malfunction, or clear the malfunction with the sound muffled by a sleeping bag (see Chapter 8).

Smell

The sense of smell is under-used by most of us. However, professional hunters learn to hunt in the dark by following odor trails. Man's sense of smell is less than 10 percent as powerful as the dog's, but it is usually used to only about 2 percent of its potential.

Different diets produce different characteristic odors in humans. For example, a heavy meat diet produces a different odor than a principally vegetarian diet. Once you become accustomed to the enemy's characteristic odor, it is easy to detect and differentiate at night, even from a distance. This may help you avoid ambushes. Practice improves this sense and builds confidence.

To use your sense of smell at night, face 45 degrees toward the wind, relax and breathe normally but take frequent sharp sniffs. Think about smell; try not to think in words or pictures, but keep your mind blank and concentrate your awareness on odor. Warning: This is not safe to do while you are responsible for security on a sector; use it only while a buddy is alert for visual and audible signs that you might miss.

Other Senses

A sense-of-direction is primarily a matter of observing details (more about this in Chapter 4). A sense of the presence of others may also be due to being aware, subconsciously, of details. Both of these senses can be improved through directed practice (how much the sense is improved, though, depends on what is measured and by whom; it is difficult to quantify from current data).

FATIGUE

Night operations are often part of continuous operations, and fatigue can seem overwhelming at night. Civilians, such as police on revolving shifts, take about ten days to become efficient on a new shift; however, up to forty-five days later, they can lose their entire adjustment to evening work if they are confronted with personal or emotional trauma; a serious problem within forty-five days may degrade their adjustment to the night shift. Thus, it's helpful to keep your men working the same shifts, if possible, and to expect temporary loss of night-fighting fitness immediately after your men experience personal or professional stress.

Sleep is hardest to resist between 0200 hours and dawn. This is a good time to attack, a good time to "stand-to," and a good time for mild exercise or stimulants (coffee). Heavy exercise just increases fatigue and sleepiness, though, and excessive caffeine results in a short "up" period followed by a strong "down." Don't let your men overdo the coffee or use "diet pills" to stay awake. Alcohol is a depressant and makes resisting sleep more difficult.

The ability to reason and to discriminate is most severely affected by sleep deprivation. Thus, leaders, computer operators, and order writers are most affected.

Fatigue Management

The effects of fatigue can be planned for and minimized for a given situation.

- A four-hour-on, four-hour-off schedule works well, and a two-on and four-off works well in inclement weather, but other schedules can be used to good effect also. No one schedule is best for all soldiers, but a specific one might work best for most of the soldiers in your particular unit. Experiment with different fatigue-management schedules to find which works best for you and your men.

- Be sure men sleep or rest during at least part of each off-shift period.

- Vary tasks. If men are cross-trained, they can rotate through various duties to reduce tedium-induced errors.

- Double-man each job requiring discrimination: FDC (fire direction center), writing/formulating orders, etc.

- Order of priority for sleep should be decided in terms of seriousness of errors, complexity of tasks, and tedium of duties. Leaders and technicians might be rated a priority of 1–2 in this system. For example, if someone has to miss sleep to check a position, the leader may make one check, his assistant two checks, and a crew chief three checks; the leader would therefore get the most sleep, since he makes the most serious decisions and processes the most complex information. In a platoon, the platoon leader, squad leader, and team leader may well be rated a priority of 1–2–3 in terms of seriousness of error.

- Some persons are more efficient early in their awake cycle, others later on. Try to capitalize on the decision-makers' best times for critical tasks—this should be planned, not given lip service. This factor may relate to circadian rhythms (the body's "biological clock"). We know that conscious effort has only moderate effect on changing a "day" person to a "night" person, but conscious effort has a better effect on changing night persons to day persons.

- Carbohydrates (sugar) require thirty-plus minutes to reach the blood stream, so think about scheduling meals/snacks at least thirty minutes before anticipated action or critical awake times.

• Visit your men frequently and try to keep them together in buddy teams.

• Body temperature normally rises in the A.M., drops in the P.M. This temperature decrease may lead to sudden sleep in a fatigued individual. Mild exercise raises the body temperature, leading to increased alertness.

• Normal awakening/sleeping time adjusts about one hour per day under non-stressed conditions. Thus, men changing to a reverse cycle may require about twelve days to become fully efficient on their new schedule. The effect of stress on the rate of change isn't evident. Give your men time to adjust to a new sleep/awake cycle.

Jet Lag

Travel through several time zones quickly leads to temporary disorientation because the body cycles remain on the old time schedule. These cycles start to catch up in about three days. Try to change to the time schedule in your area of operation (AO) before leaving. Set up new eating and sleeping times prior to your departure. Then, eat lightly or not at all prior to the flight, and your men will be partly "pre-adjusted" to the new AO and less affected by jet lag.

USING PHYSICAL AND PSYCHOLOGICAL FACTORS FOR COMBAT ADVANTAGE

• In the defense, visit men, make prompt decisions, refrain from expressions of personal insecurity or reservations. Use buddy teams in positions. When it is possible, maintain a schedule—something men can believe in. Emphasize "esprit de corps": "best squad, best company"; "light fighters can do anything"; "we're all in this together and we'll come back together"; etc.

• Maintain the spirit of attack. Advance, inches if necessary, but advance.

• Keep your machine guns (MGs) quiet when they are not needed for firepower. Unnecessary fire draws attention (and counterfire) and reduces the surprise of later sudden automatic-weapon (AW) fire.

• Act decisively and keep your men informed. Don't just react, or wait for something so you can react.

• It is sometimes useful not to return fire and not to yell during an attack. You won't give away your positions, your enemy will be less sure of himself, and you will appear controlled and invincible. Often, however, a screaming attack at night is unnerving.

• Night attacks are very useful for light fighters. According to Japanese army doctrine during WWII, "Night attacks will frequently attain difficult, limited objectives with comparatively few casualties. Attacking troops must be highly trained and imbued with the determination to close with the enemy and destroy him with the bayonet."

• During night battle, each subunit and each man needs a specific objective to counter indecision and lack of confidence.

• Practice the hardest tasks during the darkest hours. Thus, day habits will help, not hinder, your squad while it is operating in partial light, i.e., moonlight.

• In the dead of night sound carries better than at dawn or dusk. Thus, attack at dawn to take advantage of muffled sound; have the unit ready at dawn or just past sundown for this reason.

- Don't neglect calls of nature; empty bowels before starting on night operations.

- Sleep during allotted periods; rest if you cannot sleep.

- Give your men the results of what they did, or tell them how well they did. MILES (a training device whereby a light placed on a man or tank is activated to indicate a "kill" during training) is especially useful for immediate evaluation of tactics and movement during training.

- "The weightiest consideration in night movements is mental, for the mind is agitated. When the mind is not calm, there is uncertainty. One cannot distinguish between fact and fancy. . . . At night, suggestion is easy: Confusion, false reports and mistakes spread rapidly to those nearby," Japanese training manual, WWII.

2. Night Operational Techniques

THE NEED FOR STANDARDIZATION

When we're fatigued and/or under stress, we tend to rely on habits to decide what to do. Thus it is vital to develop and practice effective techniques until they become habits. If we take short cuts, explaining to our men "in combat, we wouldn't do this," we are teaching our men to fail in combat. The more precisely we practice, the better we'll fight.

Newly awakened soldiers, especially, are often not thinking clearly. The stress of incoming rounds or an imminent firefight could add so much confusion that your soldiers move out without essential equipment or with their clothing and gear in disarray. They'll soon be suffering from exposure or stopping to repack. It pays to teach your men habits that will carry them through confusion and sleepiness. Develop a "dressing SOP" (standard operating procedure) and use the knowledge to develop your unit's field pack, combat equipment, and campsite techniques. Teach your men to dress and pack the same way every time, and then, practice until getting ready to move, quickly and silently, is a firm habit. Remember, "cutting them some slack 'cause they're tired" isn't doing your men a favor.

The exact order of dressing and packing isn't critical, but *consistency* is. If your soldiers each pack the same way, buddies can help the slower men and the wounded. In addition, when soldiers dress and pack the same way every time, they can find what they need quickly by feel. If necessary, a messenger can return to the cache site and quickly collect a needed item from each pack. For example: "We're going to be here for a few extra hours and we'll have a meal here. Charlie, go down this ridge to the stream, turn right, and go about 400 meters to the cache site. Tell security that we'll be here at least another three hours and collect the rations."

After dressing and packing, your unit can move into their positions or into formation without further instruction. The more orders you have to give, the more chance of creating confusion or compromising your location. Select a standard formation and arrangement of NVDs and weapons. Modify your standard formation to correspond to your actual or anticipated mission, but be sure your men know what formation will be "standard" for the mission. Rehearse it—don't wait until you are forming up to tell them.

One technique, used by some elite units, was "borrowed" from professional football. The order to pack and be ready to move is given in regular sequence or with a prearranged code and is passed to each man as he is awakened. For example, "Moving out in two minutes, column toward five o'clock with guns two and ten, grenades five, and goggles forward and right."

In this example, each soldier as he was awakened, learned that his unit was to form a column in two minutes; that the column would form as if it were to move toward the right and rear from the base direction (toward five o'clock); that the NVDs would be with the point (the initial person to enter an area and provide frontal security), and the right flank security; that automatic weapons would face the front of the column; and that the grenadiers would be near the rear.

To insure accuracy, each soldier repeats the instruction to confirm his understanding, then wakes his buddy neighbor, gives the order *exactly* as he received it, and listens to the "briefback" to insure that what he meant was exactly what was heard. Two minutes later, the unit is formed and watching its sector of responsibility, while the leader inspects and corrects and his assistant insures the site has been sterilized.

The small unit needs to establish SOPs covering the most frequent or anticipated events that can happen at night. For example, what immediate action the unit or patrol takes to avoid a high flare, ground flare, or mine. These immediate action SOPs permit the unit to take rapid action as one man, cut down exposure to enemy observation and fire, and aid in quickly regrouping the unit. (Reaction to high flares is often an immediate prone position to eliminate shadows; reaction to a ground flare might be to move immediately out of the light: "five o'clock, 100 meters, now!"

NEED FOR SOME LIGHT AT NIGHT

Many routine duties are possible, even easy, in the dark if they are practiced. One very important responsibility that can largely be performed in the dark or with the aid of the light from a compass is first aid. The essentials to check for are *breathing,* which we can hear and feel, *serious bleeding,* which we can often feel and see in dim light, and *shock.* Artificial respiration, direct pressure, bandaging and splinting, insuring warmth, and elevating the feet can all be done by feel–if practiced in the dark.

Some tasks require some light, but light can confuse a soldier who isn't familiar with its effects. For example, blue light causes all blue markings on a standard map to disappear, and red light causes the red markings to disappear. One solution is to copy the applicable portion on a duplicator. The blues will appear black, but will be harder to identify, so practice is needed. (Water features can be hand marked with jagged lines.) Blood, seen in the light of a compass, filtered flashlight, or chemlite sometimes appears to be tar or molasses, so first-aid practice with training aids is needed.

Radio telephone operators (RTOs) or someone close to the map readers should have a poncho handy to make a "light tent" for map checks. And, strip maps (see Chapter 4) can be drawn on acetate over a luminous-taped background to reduce the need for lights. The luminous mapboard must be covered, of course, when not in use. It can be used for signaling, too (see Chapter 3).

Dials and scales can usually be read by the light of a compass. However, if you stare directly at the markings, they may disappear, so practice is again the answer. Because this skill is critical, have your men practice reading scales by the light of a compass or luminous tape, using scanning.

Soldiers must become familiar with the "feel" of their equipment in the dark and confident of their ability to work with the various pieces of equipment during darkness (see Chapter 8).

NIGHT MOVEMENT

General

Movement at night is characterized by reduced speed and the need for silence. New soldiers often rush at night, taking large steps and pushing off from their rear toe and dragging their rear foot forward before raising it. They stare at the ground in front of them. An experienced night fighter moves slowly with small, high steps, feeling carefully ahead before shifting his full weight onto his foot and at the same time visually scanning the sector.

Each soldier should try to step in the footsteps of the soldier ahead because that small space has been "cleared" (no twigs, rolling rocks, or mines). This will help conceal the size of the unit from the enemy. Bushes and branches should be carefully pushed aside and replaced—not broken off. Broken branches leave white spots on the sides of the trees or shrubs which signal the unit's presence in the area; these white marks are very visible at night.

Night Walking

When we walk on a sidewalk, we thrust our foot well forward, straightening our leg before our heel hits the ground. Then we kick our weight forward and swing the rear foot forward. This technique uses our calf-muscles to propel us fairly rapidly over smooth ground.

But night fighters use their thigh and buttock muscles to move slowly across rough terrain. When their large muscles are trained and they habitually take short, careful steps, cross-terrain travel becomes as easy and natural as walking down a sidewalk.

Looking ahead, slowly lift your left foot to nearly knee height, balancing on your right, and then ease it forward, feeling for twigs and trip wires. Point your toe down about six inches in front of you and feel for the ground with the outside of the toe of your boot. Settle your toe on the ground, then your heel, quietly, feeling for twigs or loose rocks. When you are confident of solid, quiet footing, slowly roll your weight forward. Hesitate and then begin lifting your left foot (see Figure 4).

Figure 4. Night walking. Place toe down first, slowly and gently.

Teach your men to feel for the ground with their toes. The balanced, smooth walk used at night will reduce tripping over roots and rocks as well as reduce noise. And, once your men are used to walking at night using the larger muscles, they'll travel farther before getting tired. Balance will be much easier if they scan the horizon with their eyes instead of staring toward their feet. And scanning ahead helps one to stay oriented and to detect the light, contrast, and motion that could signal an enemy ambush.

Stalking

Learn to stalk in a crouch with your eyes on your quarry. Stalking is best described as night walking in a crouch, very slowly. When you are close to your quarry, squint to avoid attracting attention to your eyes. And be sure you freeze when the quarry looks toward you; balanced or not, stalkers must not move when they might be under observation. A stalker's outline can blend with shadows and he can be free of glare and contrast, but his slightest movement can give him away. Train to instinctively move during distractions such as gusts of wind, a burst of weapon fire, or any other sudden event.

Stalking a sentry is often easiest during a shift change or while a patrol is passing through the sentry's position. Sentries tend to watch the patrol and talk to their relief instead of watching their sectors during these times. Also, sentries tend to be least alert during the period 0100 hours to 0300 hours.

Often the final stalk is made by crawling, since looking from low to high is more effective and because crawling often best utilizes low concealment.

Crawling

There are several methods of crawling. The fastest crawl is accomplished by pressing the arm and foot on the same side of the body against the ground to pull or push forward. This is a relatively noisy method, however.

A quieter method of crawling involves pressing down an arm and a foot on opposite sides and resting on one hip. For instance, your men can pull themselves forward with their left hand and right foot, resting on their left hip while they reach forward again. This is slower, but quieter than the first method.

The slowest and quietest crawling technique is to use the toes and elbows to lift, then propel the torso forward slowly, a few inches at a time.

While using any crawling technique to stalk, you should feel for twigs and other noisemakers and remove them or go around them. You may have to smother a twig with your clothing before leaning on it, to muffle the sound of it breaking.

Your men can pad their knees to make crawling more comfortable. Be sure, however, that the tape or ties don't tightly encircle the legs and cut off circulation.

All of these crawling techniques feel awkward at first. Fortunately, light fighters can practice them almost anywhere and any time.

Silent Night Attacks

Light fighters can use silent intimidation during night attacks. The enemy defenders will tend to get nervous and lose confidence when attacked or probed methodically. If they cannot determine the size of the attacking force, they may exaggerate it in their minds to an overwhelming size.

Once among the enemy, light fighters with bayonets fixed have a psychological advantage. Also, light fighters are relatively safe from enemy indirect fires while they are among the enemy positions. Finally, defenders often hesitate to shoot at shadows where their neighbors might be, but the light fighters can freely engage targets with grenades and bayonets. If the defenders are not well-trained, they may fire indiscriminately at shadows and sounds, giving their positions away.

Prone Positioning

Move into a prone position by tucking your rifle under one arm and feeling the space in front of you with your other hand while you slowly squat. You should move slowly so you can freeze in place, to avoid breaking twigs or tugging on trip wires. Slowly feel for the ground with your free hand. Support your weight on that hand and the opposite knee and then extend your free leg toward the rear. You will learn to roll quietly onto your extended leg. Extend the other leg and roll back into position without a sound. If you feel a twig or loose rocks, you can stop, feel, and remove the noisemaker, then continue rolling into place (see Figure 5).

Figure 5. Moving into prone position at night.

NIGHT POSITIONS

Digging In

The following is a description of one way of digging positions in the dark. Be sure to remember that you will need alternate positions if the primary ones are discovered. Your SOP might include moving to alternate positions automatically at night in case your primary positions have been discovered during the day.

After placing each buddy team on the terrain exactly where you want its defensive position, your soldiers construct their positions in the following manner. Facing its sector of responsibility, the team should mark off its position on the ground with a strip of tape or rocks and sticks. The men should start digging at their feet and then dig away from that spot. They shouldn't drop or throw the dirt, but put it down quietly where it belongs. They should not strike the ground, but press their shovels in and use the pick only if absolutely necessary. If for any reason, such as an enemy attack, they stop digging, they should place the tools carefully where their rifles were lying so they can find them again quickly. When the shifts change, the team should hand its tools to the relief without conversation.

The normal error made by new soldiers "digging in" in the dark is making positions too narrow. In addition, early in training, a lot of time can be lost when tools are misplaced during a break or change of shift.

Defense Discipline

Enforcing rules is a major part of ensuring discipline. If you ignore violations, soon your base will be a bubble of conversations and cigarettes being lighted and puffed, and the site will be filled with partly hidden trash, cut bushes, overturned rocks, and scratched trees.

If smoking is allowed, lighting the cigarette should always be done in a shelter. A poncho teepee takes about fifteen seconds to make and hides the light fairly well. The burning end should be continuously concealed in a cupped hand. Small campfires, built well below ground level and away from foliage or rocks that reflect light, are hard to see from a distance and are just as warm as a bonfire. However, the smoke from cigarettes or fires is easy to detect.

Bushes can be propped or braced out of the way, and rocks kept ready to return to their initial location. If your men must move around, you should establish and enforce the use of trails under trees or cliffs instead of letting them crush vegetation along multiple shortcuts. Trash can be taken away or buried deeply in a concealed location.

These concealment techniques reduce your enemy's ability to find you and help to keep him unsure of your numbers.

ADDITIONAL HINTS AND TIPS

Whispering

Soldiers can whisper by letting half of their breath out, and then whispering with the remaining lower-pressure breath. Once you are discovered, however, it is best to use a normal speaking voice to give orders so as to insure clarity and instill confidence.

Resupply

Consider using a "fire brigade" or human chain to resupply a position. Form a chain and pass all of the packages hand-to-hand, then move the squad up and pass the packages again until you finally reach the base. If you simply carry part of the load all of the way and then go back for more, you may make much more noise and disturb more ground and vegetation.

When you are close to the enemy, or in the open at night, try having your men lie down in line, on their backs, with each man's head between the next's knees. Then they can pass the materials by reaching down to their knees to receive a package and then handing it to the next man over their head.

Stream Crossing

Move slowly and allow gravel or rocks to settle beneath each foot before putting weight on it. Test each foothold before putting your weight on it to avoid making noise by slipping on slick or loose rocks.

Your squad's SOP should specify a procedure for crossing obstacles; nearly always, it will involve securing the far side and the flanks before crossing.

Formations for Movement

A memory aid to help remember what to consider when you choose a formation is NOSE: *N*earness (of enemy), *O*bstacles, *S*ecurity (what do I need here?), and *E*xposure (to observation, fire, or tracking). Night vision devices and at least one automatic weapon should be near the front while moving; however, AWs normally don't accompany the point. Instead, they closely follow, probably with the squad leader.

Hints

- Don't let your men comment if they fall; a voice carries farther than a thump and identifies the source of the thump. Also, continuous griping can affect the morale of your unit.

- Don't run at night, if possible. If you must run, lift the feet extra high.

- Push bushes and tall grass aside with the foot before feeling for the ground with the toe. In short grass, lift the feet high; in long grass, keep them low.

- Keep men informed of the reasons for halts; while they think about and look for a reason for the delay, they ignore their sectors.

- The key to successful night attacks is the simplicity of your plan and a *thorough* understanding of the plan by every one of your men.

- Have your men jump up and down before leaving so you can correct gear rattles.

- The most effective "killers," as evaluated in both combat and MILES exercises, are usually calm and quiet individuals who carefully plan what they are going to do. For example, one successful commander as he traveled always pictured possible ambushes and then planned his responses. See "Visualization" in Chapter 8.

- Patience is critical to success in night operations. During WWII, experienced Japanese soldiers crawled slowly along the entire route to their objective, then used grenades and bayonets to great advantage. The newer and less experienced troops lacked patience, moved faster, and were far less successful.

- Prior to an assault, try to remove and cache every extra piece of equipment. Take ammo, water, and grenades. Consider carrying grenades, with handles taped, and magazines (taped in pairs facing opposite directions and offset) in trouser pockets. Be sure the ammo and grenades don't rattle, though. Webgear may not be needed during an assault; and it does make noise. Develop a squad "stripped-to-assault" SOP so you won't have to explain details on every operation.

- The Kevlar helmet impedes hearing; at night try to have the point elements move and patrol in field caps or bareheaded to help detect the enemy before he detects you.

- In the defense, don't let your men wander around the area; they may be mistaken for the enemy or compromise your location, and the confusion caused by wandering soldiers is a distraction to those on guard.

- A man can often detect trip wires by holding a stick lightly, pointed about 45 degrees down, and held about two inches off the ground.

• Don't let your men bury their faces while under fire; if they watch the tracer paths, they can better describe where the fire came from and thus locate the enemy weapon. (Many armed forces light fighters may face use of green tracers.)

• Ladders (even expedient ladders), reed or cloth mats, and screening can be used to cross wire obstacles quickly. The mats or screening are thrown over the wire, which the men then run or roll over and quickly move on.

When you issue your order, have your men explain to you what they are to do so you'll know that they understand your plan. A tired, frightened man may not hear exactly what you say; in addition, since you are very familiar with your plan, you may expect him to understand something you haven't told him. A well-briefed man might reply:

> I'll cover the right flank until we get to the water tower. If we go through fog or smoke, I'll move closer to the unit so I can see the unit as we travel. We'll be traveling east for about 600 meters. If we're attacked, I'll return fire. If we disengage, I'll return to the last rally point (RP) unless you yell a clock direction; then I'll go that way, considering our direction of travel to be 12 o'clock. I'll call you, if I find trouble, by clicking two rocks three times. Is that right, Sarge?

The officer might reply: "Mostly. Remember not to fire unless I give the word or you are in really deep trouble; we don't want them to know we're in the area if we can help it."

3. Signals

Signals you use to control your squad at night may be different from those you use during daylight. Hand signals may not be visible during darkness. Also, additional signals may be necessary at night. For example, men reaching a trail junction cannot determine which branch to take by looking down the trail to see where it goes. In fact, they may not even be able to see the junction when they reach it.

The sound of radio static and field telephone ringing carries long distances at night. Further, the raised voices often used unconsciously can compromise your unit. If you must use radio or telephone, consider using a preplanned signal of clicks produced by keying the transmitter, and use earphones. Generally, radios and telephones are not used during stealth operations, but they may be useful after the squad is discovered or on the final assault.

Think through your operation before you start and ask, "Where will I need to give my squad or point man instructions?" Then plan for a silent method of telling your men what they need to know and also for another method if the first won't work. For example: "I'll want to tell the squad to move across after we have secured the flanks and the far side of the stream we have to cross. I can either signal with my compass, or click two rocks if it's foggy."

Consider, specifically, signals to control movement of your squad and to control fires.

MARKING SIGNALS

Breeches through obstacles, trail junctions, and rally points will normally be reconned and left under observation; the sentry can direct your squad onto the correct route. However, chemlites, light powder (such as soap or flour), piled rucks, a slanting stick, or even a line in the dirt can be used.

The key is to make sure the signal will be noticed. If you tell your men how far they'll travel to reach the marked point and describe the terrain, it is less likely they'll miss the junction.

You will want to mark temporary cache sites and rucksacks so they'll be easy to identify; luminous tape and colored chemlites work well. Also, consider using night vision devices and IR (infrared) strobes to signal sites to your squad. The SOP should specify the pattern of luminous tape that will be used to mark the inside frame of rucksacks. Normally, rucks should be left together in an ORP (objective rally point; the rally point just before an objective, usually an assault) so they can all be found easily. These signals can carry specific information; for example, if you cut two bars in a rations package and leave a chemlite inside, the two bars may signify the patrol base, but three bars would indicate a cache site, and one bar a trail junction. Your squad's SOP can specify what each signal means so you won't have to describe each signal for every operation. Also, a guide will be able to mark your route without detailed instructions if the code is part of your squad SOP.

Another marking you may want to plan is an assault route. A trail of a particular color of chemlites can mark the route to the objective, helping units to maintain their direction and get support or supplies to you quickly. Of course, you must tell them what to look for before you leave. A fully exposed chemlite may be too bright, though, so consider taping your markers or leaving part of the wrapper on.

SOUND SIGNALS

During an assault, surprise and stealth are usually lost at some point. After that, clarity is more important than secrecy. Consider whistle signals, at least as a backup, for your voice may not carry as well, and your words may not be understood. If your men are thoroughly briefed about what they are to do, they'll be more likely to expect a particular command and listen for it.

During WWII, airborne units used metal "clickers" or "crickets" to communicate as they moved over their objectives around the Normandy invasion sites.

Light fighters can click two pebbles together to signal if they have no "crickets." Pebbles can be clicked, scraped, or tapped on a rifle stock to give different messages. Their sound can be used as a challenge/password to identify friendly personnel as long as everyone using the signals thoroughly understands them. For example, "During tonight's patrol, two taps on the rifle are answered by rocks scraped together two times."

OTHER SIGNALS

A soldier's specific activity can signal certain information. For example, if a sentry walks to the rear of his OP (observation post), it means he has seen activity. You can also signal by example, telling your men to move out by moving forward yourself. (Be sure they pass the signal on to those who can't see you, though.)

The point man moving his compass or field hat (with luminous bars) side-to-side (making the luminous bars move horizontally) can be a signal that he has reached an obstacle or danger area, while vertical motion can signal "enemy sighted." At a greater distance, the same signals can be made by waving the hat at the top of an outstretched arm or signaling with a taped flashlight or masked chemlite.

The key to communication is understanding. You must ensure that your men know what the signals mean. Continuous practice during field exercises will help them learn your SOP signals, but it pays to question each man periodically to make sure he understands your squad's signals.

SIGNAL HINTS

- Don't rely on pyrotechnics alone; have an alternative.

- A can filled with chemlite or burning fuel-saturated sand, with a cutout facing the people you are signaling, makes a good marker. The cutout can be a specific, simple shape to communicate specific info. For example, an "X" could mean "stay where you are; the operation is delayed," but a vertical bar could mean "it's on; go to your assembly area." From a distance, however, light patterns may not be possible to distinguish without binoculars.

- Consider having signals in your SOP for "advance," "move back to last rally point," "stop," "freeze," "suspicious activity seen," leader's location, "maneuver right or left."

• If your troops are alert, small paper tags (for example, on the right side at knee height), broken branches, or strips of cloth or tape can mark a route.

• Consider marking destroyed bunkers or positions with chemlites so you'll be able to find them again (after the assault).

• The SOP should specify the filters and taping for vehicle lights and flashlights, and the chemlite colors and/or markings (X- or bar-shape cuts in the wrappers).

• Light fighters should practice delivering exact messages. Personal interpretation is natural, but can be disastrous. To train your squad, consider using messengers even if a radio is available. Also, have every message confirmed *each time it is exchanged* by having it repeated. Remember, though, that movement by a messenger in ambush sites may give away your position.

• Keep messages simple. According to Murphy's Law, "If it can be misunderstood, it will be misunderstood, at the worst possible time."

• If necessary, color code flashlights to identify the AW, grenadier, and the leader.

• Ask yourself, "What do I need to say, when? How can I best say it and maintain both security and control."

• Clarity in communication during every action is critical at night. Think about ways you can communicate more clearly.

• To indicate a direction for your men to fire during a firefight, or to create a reference point from which to direct fire, tape a chemlite to a rock (or mark the rock with chemlite "juice") and throw it in the correct direction.

• Tape a specific shape, such as a cross, on a flashlight filter, and flash it toward the soldier receiving the message. For example, "When you see a red 'X' from my flashlight, move the point forward again."

• A tug on a line or piece of WD1 wire can be an effective signal between security elements and assault or support units during an ambush. Here again, night vision devices such as PVS-5s may be useful for signaling silently.

4. Night Navigation

Navigation at night is not inherently more difficult than day navigation; it is simply different. Certainly, the difference can be intimidating if you don't make a positive effort to learn new skills. But night navigation, like night walking, can be learned by anyone who keeps his mind open to new ways of operating and practices what he has learned.

Navigation at night uses the same two basic methods as day navigation: 1) terrain association, which uses a general direction of travel and the recognition of terrain features on both the map and the ground, and 2) dead reckoning, which uses a compass direction and specific distance.

Some of the techniques used to maintain orientation during the day are even easier at night. For example, it may be relatively easy to stand near a hilltop at night and read azimuths to a lighted radio tower or a town that appears as a glow in the sky. During the day, that antenna might have been hard to see, the town hidden behind a hill mass, and your hilltop exposed to enemy fire.

You can use resection—plotting back-azimuths from two distant objects you can identify on the map and on the ground—both day and night. Your map location is at the intersection, or crossing, of the back-azimuths, regardless of the time of day or visibility.

However, most night navigation techniques require that you take extra care to remain oriented. Thus, to be able to use resection and stay oriented at night, you'll have to do a thorough map and terrain analysis (map and ground recon). Usually, the most critical navigation you'll do is finding the spot to begin an attack, so become very familiar with the terrain surrounding your ORP. Remember, too, you may be crawling to the ORP and to the objective, so picture the landforms as they'll appear from the ground.

MAP ANALYSIS

When you arrive in a new area, find out where you are on your map and orient the map to the terrain. Ask yourself questions as you analyze the local drainage or relief and try to picture the landforms shown on the map. "What landform am I standing on? How high is it? What direction does it run? What are the highest points within a day's walk north? South? East? West? How much altitude will we gain (or lose) if we travel in each direction for two hours? If I could pour water on each ridge, where would it run? If I stand on this feature, will the landforms in each direction appear higher or lower than my position? How will the same landforms appear from another feature? What is the latitude and longitude of this area?"

When you finish asking and answering questions—the same questions you might ask your subordinates to determine the thoroughness of their recon—you will be able to mentally picture the landforms around you.

In fact, you will be ready to build a sand model of the terrain, which is an excellent method of improving your map-reading skills.

After you have practiced this questioning technique you will be able to:

- Mentally picture the landforms and their silhouettes as seen from your location.

- Determine the direction of each landform and the relative steepness of each landform in your vicinity.

- Identify high ground and other key terrain.

- Identify the travel routes for foot troops and vehicles (avenues of approach) toward key terrain features in your area of operation.

- Determine what the skyline may look like in any direction from a specified location in your AO. (Silhouettes of trees or brush may alter your imagined skyline, though, so compare what you actually see when you get to your AO with what you expected to see.)

The ambiguous terms used above, such as "may look like" and "mentally picture," point out an important fact: You have to walk the ground to really know it. If you have time for a recon in each important direction, it will be time well spent. If not, your ability to move will depend upon how thoroughly you questioned yourself during your map analysis.

Continue your map recon by thinking about how you might be able to recognize the important features at night:

"If I walk up this ridge northeast, I'll have a steep canyon on my right. I probably won't be able to see the river, but I might be able to hear it since it's only about 500 meters away.

"After I cross the peak, I'll be in a saddle, but the trees may make it look like I'm in a hole. If I look around, though, after I pass the first hilltop, I'll find the ground slopes downward in two directions as it did on the ridge leading up to the hilltop. But the ground will slope upward in two other directions, so I'll know I'm in a saddle. If I find the ground slopes upward toward the northeast and southwest, I'll know I'm in the right saddle." (See Figure 6.)

TERRAIN ANALYSIS

As you recon the ground, look at your map and confirm or modify your guesses about its appearance. Forming a mental picture and modifying it, section by section, is easier than trying to merge all the mental images of the map and ground together. And, improving your mental image of the landforms helps fix the patterns and important features in your memory.

As you recon the ground, ask yourself questions that will help you stay oriented in the area:

"Generally, in the Northern Hemisphere, houses tend to face south, is that true here? Does the wind seem steady from one direction? What time does it shift? Are wind rows in sand or snow consistent in direction? Where do they change direction? Do farm furrows and fences tend to run in certain directions? (Furrows often run east and west unless they are on hills.) Which way do the rivers run? What features will I be able to see at night?"

When light dims, subtle differences in light and dark shades disappear, and color contrast becomes less

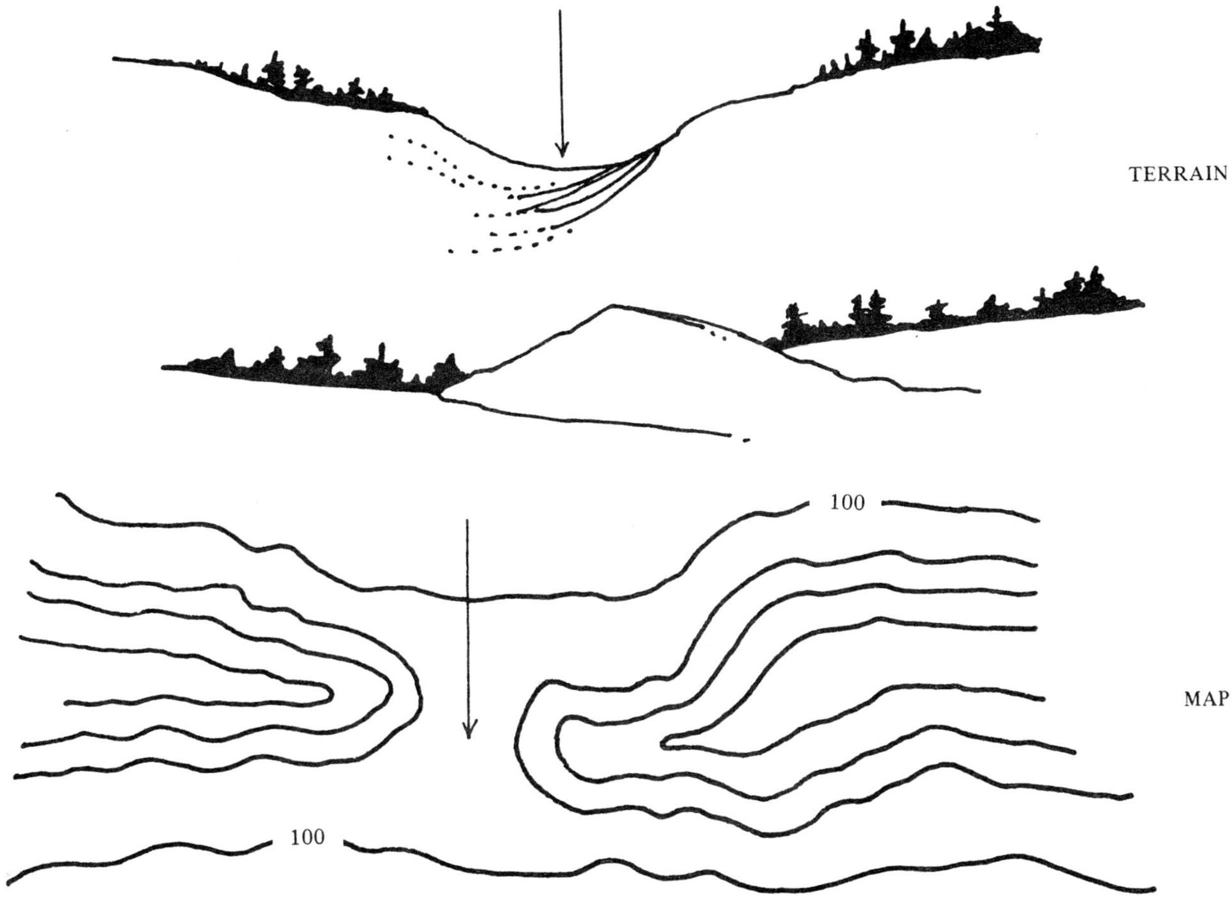

Figure 6. Saddle. Schematic drawing only.

noticeable. Thus, "a light green tree among dark ones" or "a pine at the base of a granite cliff" that you noted during a day recon may disappear, leading you to believe you are lost. But, "a lone tree on the ridge, and three rock pinnacles silhouetted to its right" might be easier to find.

The answers to the questions you ask yourself will help keep you oriented as you travel through your AO and will help you avoid gross directional errors, such as using the south-seeking, rather than the north-seeking, needle to measure an azimuth.

Some people seem to have a better "sense of direction" than others and may even say they "never get lost." Current research indicates that their ability is primarily that of noticing details. Thus, as you practice, you'll become "lost proof" yourself.

ROUTE SELECTION

In planning any route, we first consider *M*ission, *E*nemy, *T*errain, *T*roops, and *T*ime (METT-T). Occasionally, one route will be implied by the mission: "Recon the ridge." Or the route may be planned to avoid enemy troops, enable tired troops to save time, etc.

Often, though, several routes may be usable, that is, meet METT-T requirements. If you are selecting a route for night travel, your next consideration is *ease of navigation.* Finally, remember that night travel is strenuous and all night operations are stressful for inexperienced soldiers. Thus, if all other factors are equal, choose the easiest route.

Keeping METT-T in mind, look for routes to your objective. Pay particular attention to those that will be easy to follow in the dark: routes that parallel a land feature (ridge, river, or road); routes that require crossing a series of ridges at the same angle; routes directed toward easily recognized high ground.

Select the easiest route to follow that meets METT-T requirements. If several routes are similar in ease of navigation, analyze them further. Otherwise, analyze only the best choice for: *O*bservation and fields of fire, *C*over and concealment, *O*bstacles, *K*ey terrain, and *A*venues of approach (OCOKA). See if you can adjust your route to use OCOKA factors to your advantage, but keep it easy to navigate.

Your METT-T may make one OCOKA factor particularly important. For example, if your mission is to find and fix a retreating enemy force, key terrain and avenues may be critical factors. In any case, ask yourself questions about each factor.

Observation and fields of fire: "Where is observation best? Worst? Can I adjust the route to improve observation?"

Cover and concealment is worst where? "Can I improve it? Will I need help (smoke, distraction)? If I'm spotted here, where can I go quickly?"

Obstacles are features that impede. "I usually cross a river by securing the flanks, sending security to secure the far side, then bringing the squad across using the best method for the mission and situation (line, column, or group). Does this river require special crossing techniques? I'll be vulnerable while crossing so I need to find the darkest or most hidden spot. Will I need to breach this mine field or can I get around it? Where will I position security while we clear a path through the wire?"

Key terrain is high ground or tactically important features. "Key terrain along my route offers an advantage, but it may contain enemy forces or be under intense observation. If I pass key terrain, I may be crossing avenues of approach; what can I expect on these avenues? How will I maneuver if I meet a force here?" Mentally picture both enemy actions and your own. See Chapter 8 for a description of visualization techniques used by some successful sports teams and elite military units.

Avenues of approach are routes to and from an important terrain feature, key terrain, or fixed position. A major reason for identifying avenues is to determine where you are most likely to meet another force, friendly or enemy, while you travel.

ROUTE LEGS

Once you have identified the most easily navigated route that meets your METT-T requirements and have adjusted it to best use OCOKA factors, how are you going to follow it?

To plan, first identify each point where you will change direction. You'll want to be sure to find these "checkpoints" on the ground, so describe each of them in your mind:

"Let's see, I'm going to move from here to the vicinity of these coordinates by 0400 hours and be ready to move again by 0800 hours. There are small enemy patrols in the area, and we're on the edge of the battalion's area so we could find lost friendlies in here, too.

"We are to destroy enemy troops or to fix a larger enemy force and get help if we find one; all vehicles in the area will be enemy.

"The troops are in good shape and the terrain in the area isn't bad—it won't affect our mission.

"We have time to engage some targets of opportunity and that's part of our mission. We'll have to find them, though, without getting lost; we can't afford to wander around in the valley much past 0300 hours.

"On the first leg, we'll travel along this streambed, generally northeast and parallel to a ridge on our right, for about 250 meters. When we reach the footbridge, where the stream bends sharply left, we'll get on the finger of a broad ridge that points at the bridge from the west.

"Then we'll follow a heading of 290 degrees magnetic—corrected for the declination—for 400 meters. We'll go over a small ridge and a gulch as we travel, and end up at the base of a large ridge.

"We'll turn left, since I offset to the right, along the ridge and find a deep gulch cutting into the ridge to our right. If we come to a stock-watering tank on our left, we've gone too far, and we'll have to go back to find the gulch.

"We'll travel up the gulch on the left side, in the shadows, since that's south and the moon rose from the southeast last night. We'll only cover about 200 meters along the south side of the gulch before we start to leave the ridge; we'll set up our base in that vicinity."

Then analyze the tactical factors. "Key terrain initially is the high ground to our right front, then the hill mass above us to our right. Major avenues for foot traffic are the gulches and, for vehicles, this broad ridge running toward key terrain. We'll want to be careful to identify any folks we see, for another friendly patrol could wander into the area. Any vehicles, though, will be enemy. We'll want to review our SOP for armor and I'll want to watch for a mechanized attack from my left as I approach on the big ridge.

"Observation and fields of fire should be good on the big ridge and through the last gulch, poor while we're on the compass leg. I'll want to watch flank security carefully here because enemy patrols might follow the little ridges toward the high ground to our right. This might be a good place for a hasty ambush. I'll talk to the lieutenant about sending somebody out this afternoon to select a site and watch the area.

"Our only obstacle is the bridge, which we don't want to cross—we're just using it as a checkpoint. If we have to come back in a hurry and use it, there's room for security above and below it. I'll want the unit to hustle when we cross, since obstacles are often covered by fire."

When your route's features are clear in your mind draw a strip map and record each leg in your notebook. Writing out each leg will fix its important points more firmly in your mind and help you find any areas in your plan that aren't completely clear. (See Figures 7a and 7b.)

Then go back over each checkpoint and ask, "How will I know if I miss this one?" Select a "catching feature" beyond your checkpoint, preferably a linear feature like a ridge, road, or river. "If I go too far along this finger, I'll come to a streambed; if so, then I'll go back and find the checkpoint."

Then look at the route segments, or legs, between the turning points and other critical points you have selected and ask, "How will I stay oriented on this leg of my route?" Explain your plan mentally.

If part of your route involves dead reckoning—following a specific compass heading for a specific distance to a checkpoint—try to select the checkpoint on a linear feature. Then plan to use either offsetting or bracketing to find the checkpoint after you find the linear feature.

In offsetting, for review, you deliberately follow a course with an error "built in." When you arrive at the feature, if you had offset to the right (increased the magnetic azimuth), for instance, you would look for the checkpoint to your left along the linear feature. (See Figure 8.)

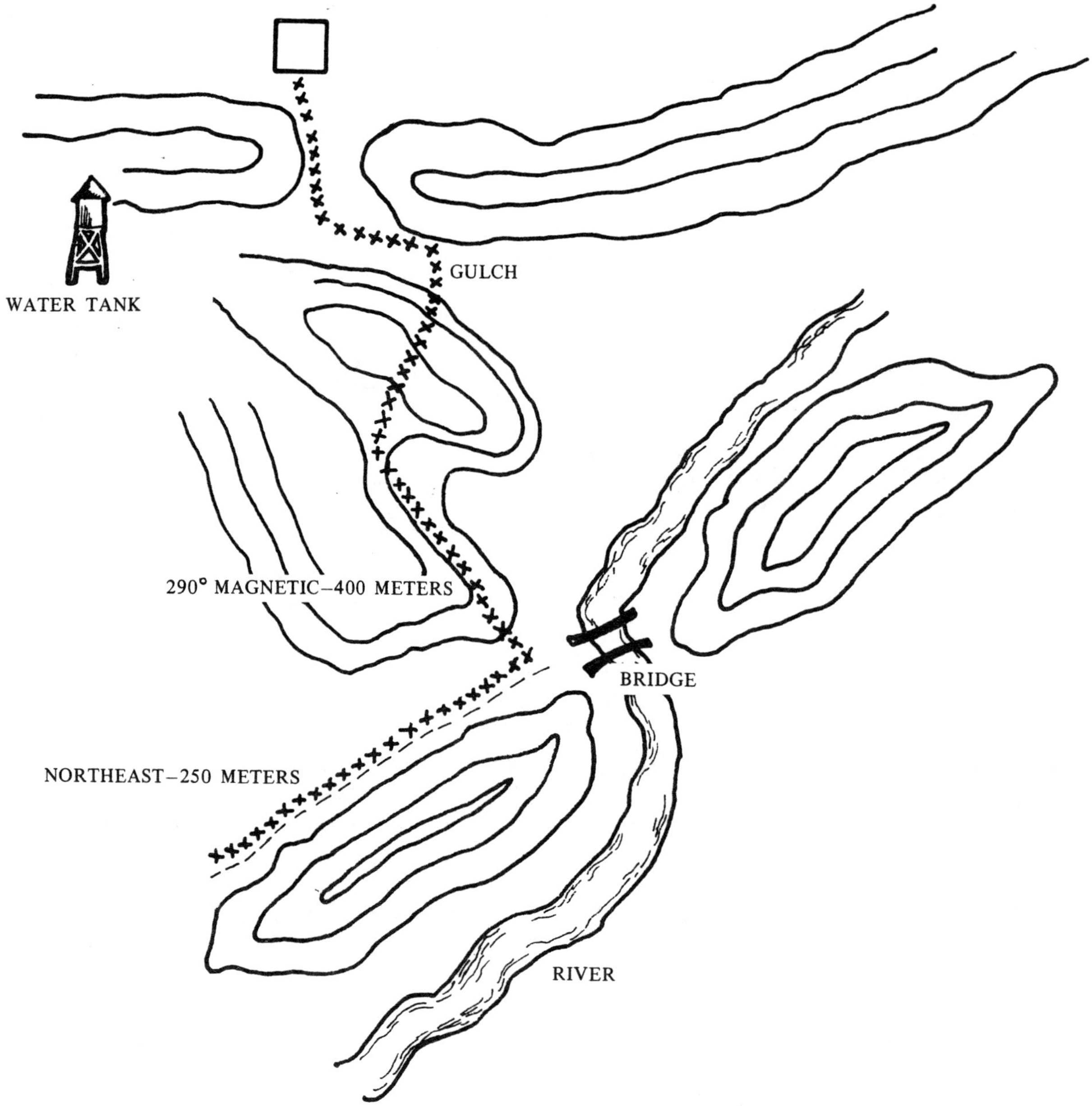

Figure 7a. Strip map. Schematic drawing only.

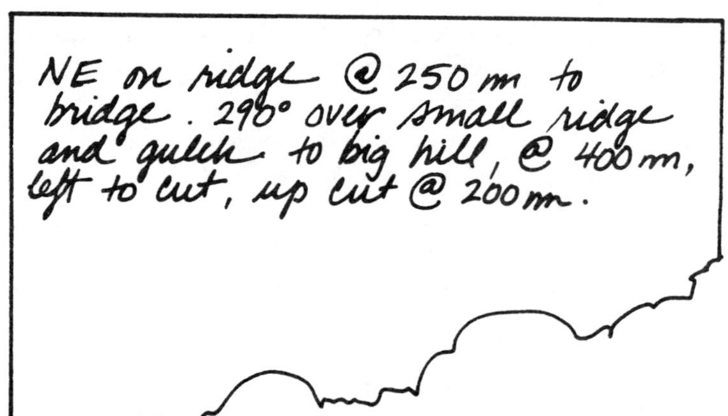

Figure 7b. Notebook entry for route.

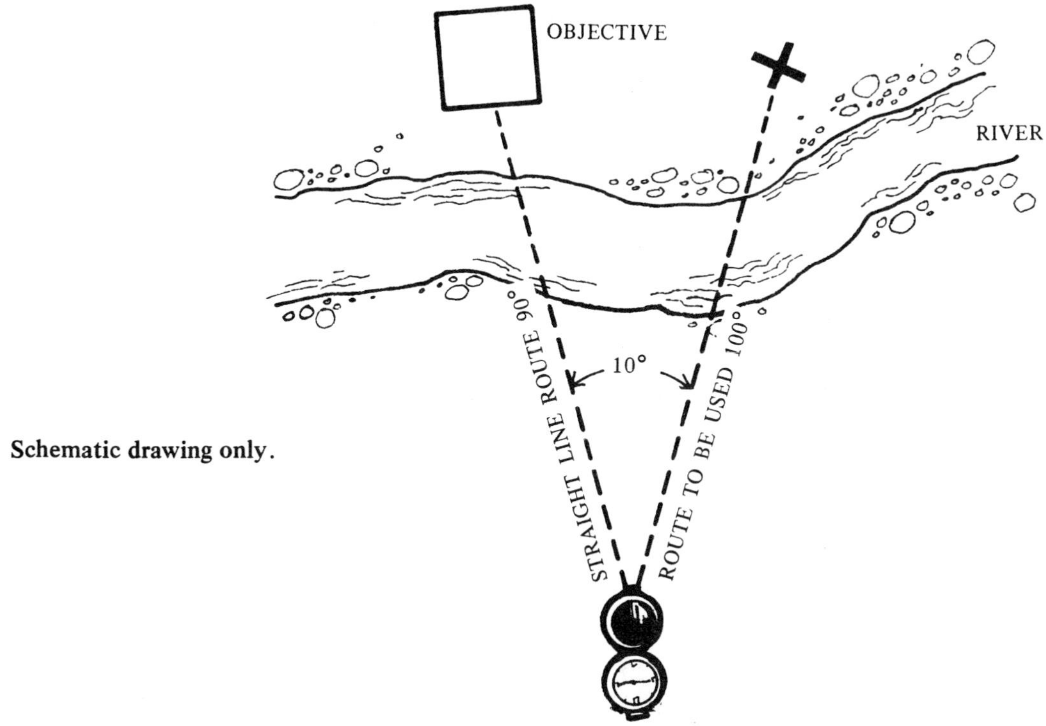

Figure 8. Offsetting. The distance from "X" to the objective will vary directly with the distance to be traveled and the number of degrees offset.

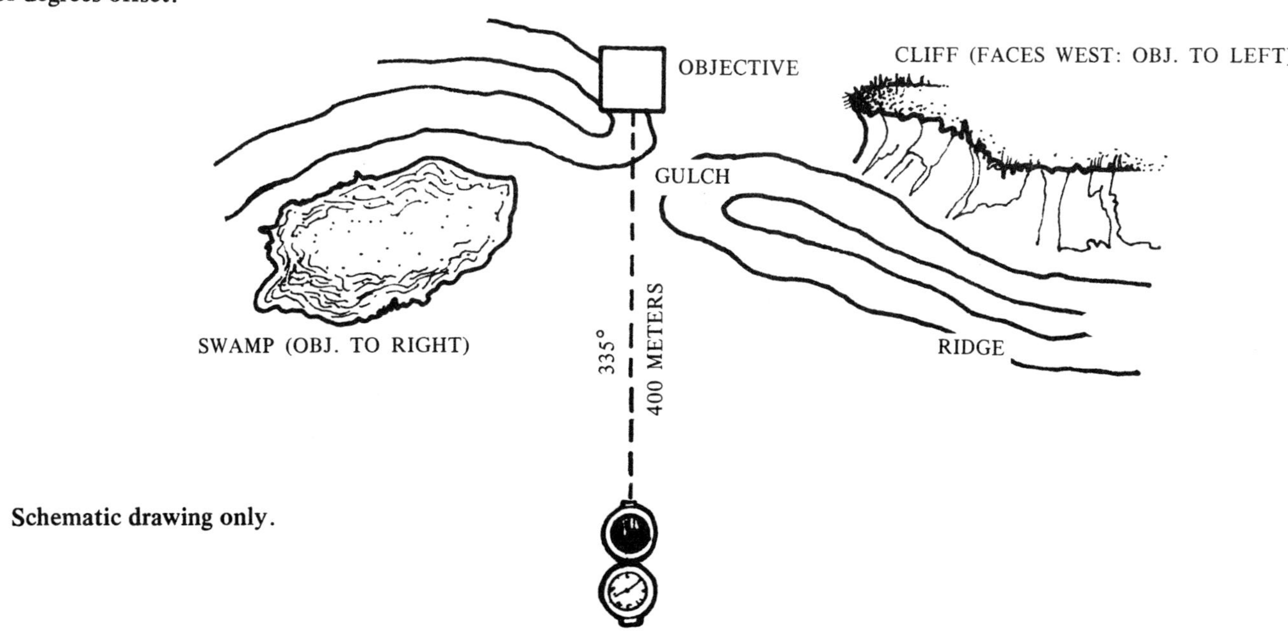

Figure 9. Bracketing.

To bracket a checkpoint, select an easy-to-identify feature on each side of the checkpoint along the linear feature. When you reach the linear feature, turn either way. You'll either find the checkpoint and proceed along the next leg, or you'll find a bracketing feature and go back, following the feature in the opposite direction until you find the checkpoint. For example: "I'll travel 400 meters at 335 degrees magnetic up to a ridge where an intermittent streambed cuts through; the stream will be in a gulch. On the ground, I'll follow a heading of 355 degrees, then turn to my left when I reach the ridge." Or: "I'll travel 400 meters at 335 degrees magnetic, then turn whichever way seems right. If I hit a cliff facing west on my ridge or swampy ground, I'll turn around and look for the gulch in the other direction along the ridge." (See Figure 9.)

If you travel a long distance by dead reckoning, your errors may accumulate. You may want to select several bracketing features in case you miss your checkpoint by a large distance. In this case, you'll have to remember which way to turn when you find a bracketing feature.

You may have to plan to "inchworm" along part of your course. For example, if fog settles while you are navigating by terrain association, you'll have to switch to dead reckoning. And, in fog or snow, or in tangled terrain, you may have to inchworm your unit all the way.

To inchworm, signal your advance party to move forward and then stop the men just before they disappear. Signal one of the men into line with your heading, signal the other men to secure the area, and then move the rest of your unit forward. Position your compass where the "marker" is standing, then send the advance party out again and repeat. Remember to keep your body in the same position each time you read the compass so your headings will be consistent.

Inchworming feels slow, but it is much faster than getting lost and starting over. Using every technique to stay oriented, and being calm and careful as you navigate, will get you close to your checkpoints. Planning "catching features" beyond the checkpoints, selecting bracketing features, and planning offsets in dead reckoning routes will make your navigation efficient.

Be sure to describe your route in words, mentally. Pretend you are instructing a not-too-bright messenger. (You may not be very sharp later, either; fatigue and lack of sleep, common in night operations, degrade your ability to think logically. Keep your instructions simple and foolproof.)

One way to record a route is by drawing a strip map as in Figure 7a. Another is to record each leg of the journey on a separate page in a notebook as a description (see Figure 7b). Whichever method you use, include checkpoints, that is, terrain features you see on the map and are sure you can identify on the ground. You'll use the checkpoints to navigate, but you don't have to plan on crossing each checkpoint. You'll simply identify the checkpoints as you travel to confirm your location.

Record the "catching features" you have selected to "catch" you if you go too far. Also record the significant avenues of approach and the obstacles that affect your travel on each leg.

Your map analysis, or map recon, will probably raise as many questions as it answers, so the next step is a terrain analysis—a route recon. If you have developed a mental picture of each leg of the route, you will be able to modify the picture as you recon, which is easier than developing an understanding of the route from scratch while you recon. If you don't have time for a route recon, the mental picture you develop during your map recon will be the key to successfully navigating the route at night.

Hints for route recon:

- Use your lensatic compass frequently.

- Azimuths readings, as well as call signs, can be written in grease pencil on acetate and mounted over luminous tape.

- The size and location of the moon may dictate which sides of ridges are in shadow (thus best for concealment). Keep track of when and where the moon rises and sets. Day-to-day variation is slight so you can plan tonight's operation anticipating approximately the same moonlight and direction you observed the previous night.

- Warning: Do not record friendly military positions or any other information of tactical value as you record the features of your route, for your map may be compromised while you are in unfriendly territory.

TERRAIN ANALYSIS DURING ROUTE RECON

Check your choice of methods of navigating on each leg. Will dead reckoning be necessary or can you use a general direction and terrain features? Record in your notebook your headings and directions and the terrain features you choose as guides.

As you recon your route, look for terrain features you will be able to identify at night. You might not be able to walk the exact route, but try to picture, especially, what the skyline will look like at night from the route. And ask yourself more questions.

"Where could we get confused and how will I be able to get reoriented? Can I find this checkpoint in the dark? Where would I set up an ambush if I were the enemy? Where is the best rally point if we have to break contact here? Where will I want my security teams when we cross the stream? Go up this brushy hill? Cross the ridge? Where will I want my automatic weapons on this leg of the route? How far apart can we travel without losing control?"

If the situation allows you to travel the exact route, step off the distances and modify your notes. Also, consider marking or manning critical turns and danger areas. For example, if one location looks natural for an ambush, consider leaving a soldier to watch it until your unit arrives. Or, mark a forked ridge with a slanting log to ensure you'll be able to follow the correct fork at night. See the description of visualization in Chapter 8.

Look for terrain features that you can use as guides as you travel and features that will tell you if you are off course. And look for expedients to help you stay oriented.

"Does the wind seem constant? Which way do the rivers run? Do houses tend to face east and south? Which way do fences and farm furrows tend to run? Will the direction of snowdrifts or sand ripples help me stay oriented? Are there any roads here? Power lines? Where will the moon rise? When?"

Make a "got lost" plan and look for locations of the orientation aids you need.

"GOT LOST" PLAN

Staying oriented is an acquired skill that you can learn by practicing. People who "have a good sense of direction" or "never get lost" actually have no special sixth sense. They simply have learned to pay attention to details, perhaps subconsciously. If you consciously practice noticing details by asking yourself questions about terrain, you will soon begin to notice those details without thinking about them directly.

There are two degrees of "lost." If you escape from a distant POW camp, you may be seriously lost and need to use the information in the next section. But, if fatigue, the confusion of a firefight, or loss of confidence makes you "feel lost," you're just disoriented. You'll be back in business quickly if you don't panic.

Since anyone, even light fighters, can get disoriented, it's wise to plan ahead of time how you'll get reoriented. First of all, what about some extra checkpoints? As you do your recons, pick out features you can use to get your bearings. Ask yourself, "What can I look at to see where I am?" Perhaps your AO has a large hill mass in the northeast corner and a ridge with a "notch" in it running from the hill mass to a river along the western flank. You know that the vegetation and sound of the river will keep you oriented while you're in the western area at night. If you plan for on-call weapon fire at a spot on the hill mass and for two or three spots on the ridge, you'll have your checkpoints.

Select points you can see from a distance. Then, if you are disoriented at night, you may be able to call for fire on a point that you think you'll be able to see; read the azimuth to the point and plot the back-azimuth on your map. Your location is where the back-azimuth intersects the terrain feature you are standing on (such as a ridge). If you need another "fix," call for fire on another point and use simple resection.

Warning: Select points for on-call fire on terrain features different from the one you are standing on. For example, if you're standing on a hilltop, but are not sure exactly where, don't call for fire on any point on another hilltop!

Preplanning for orientation fires is best, but in an emergency you may be able to call for fires on a map-coordinate position. Know what features are useful for orientation at night *before* you start. Rounds, even white phosphorus rounds, may be useless as orientation aids if they are dropped out of sight in a gulch.

You may want to plan—and coordinate—heavy machine-gun fire, too. A machine gun firing a burst of tracers down your left boundary may get you reoriented. Also, you might be able to coordinate a preplanned "light fix," that is, have a searchlight at a known location turned on briefly—but not aimed at you. If the light is IR, you'll need an appropriate night vision device to see it, of course.

Thus, if the tactical situation permits, you can resect from two preplanned fires or lights to determine your position, or you can find your boundaries by calling for lights or machine-gun fire. The key to using navigation aids is planning for them as you recon; ask yourself which terrain features you would like to be able to identify in the dark and plan for on-call fire on these features.

Although you may not always be able to use planned machine-gun fire and lights, asking the questions and planning for orientation aids will help you remain oriented. And the details you notice will solve a lot of disorientation problems. For example: "Let's see, there was a hill mass ahead of us when the firefight started, and the North Star was to our right. The wind was behind us, from the east. We moved left to break contact so we are about 500 meters south of there, and the wind is in my face so I'm looking east. The hill mass we were going toward is behind me and to the left."

Preplanned reorientation aids are useful in small areas, too, to avoid searching for a position. That is, you can leave a guide or marker near a recognizable feature before you depart. For example: "Have the assistant gunner meet me by that lone pine tree on the ridge at 2100 hours. I'll have him lead me into your security position."

Some other helpful hints are:

- A length of WD1-gauge wire from a prominent terrain feature to an ORP will help you find a well-camouflaged ORP in the dark.

- The greatest night-navigation aid is a guide; use guides whenever possible. Guides must be placed and instructed but that time is well-invested, for you'll be able to move your squad more quickly, more quietly, and with more security.

- Consider making a plan to occupy a position by "hooking in," going past it to check for an ambush and then entering it from the far side to ambush any enemy following you. Always secure positions before moving into them.

- When crossing a depression or woods, pick an exit point landmark when you head in, because you are likely to deviate as you go through and will need a landmark to get back on the original route after emerging.

To summarize, when you are disoriented, stop as soon as tactically possible and get control of yourself; panic can make you believe that you are lost. Think back to the last checkpoint you were sure of and mentally review what you did after leaving: "Which direction did I travel? How far? What terrain features did I notice?"

Then predict where you might be and check the map and the ground to confirm your guess. If you still are unsure, and if you can, call for your preplanned orientation aids (lights, flares) and use resection to plot your map location; confirm that location by identifying the nearby terrain features.

If the tactical situation doesn't permit using the aids, mark your present location and travel around it in increasingly larger "boxes" until you find your checkpoint or a feature you recognize. To box your present location, travel–by compass–a specific direction as far as you can while keeping the marker in sight. Turn exactly 90 degrees (either right or left, but be consistent from here on; turn the same way each time). Travel the same distance, turn and travel the same distance plus an increment large enough to make progress but not large enough to allow you to miss the checkpoint you're trying to find. For example, you would add less distance to the third leg of your search if you were looking for a foxhole around bushy hills than if you were looking for a water tower on gentle, grassy slopes.

The third leg plus the increment is your new distance; travel that distance, turn, and walk the new distance before turning. On the third leg, again increase the distance by the same increment. Continue to search along the sides of the "box," adding an increment to the distance you will travel on each third leg. Eventually you will find either a point you recognize or your objective (see Figure 10).

During particularly low visibility you may have to "inchworm" your unit along the legs of your route or along the legs of your search pattern.

Movement at night is slow; if you try to rush, you'll be more likely to get lost as well as to compromise your unit with the noise that is made. Strive for smooth, quiet, methodical travel at night, for it is faster and safer in the long run.

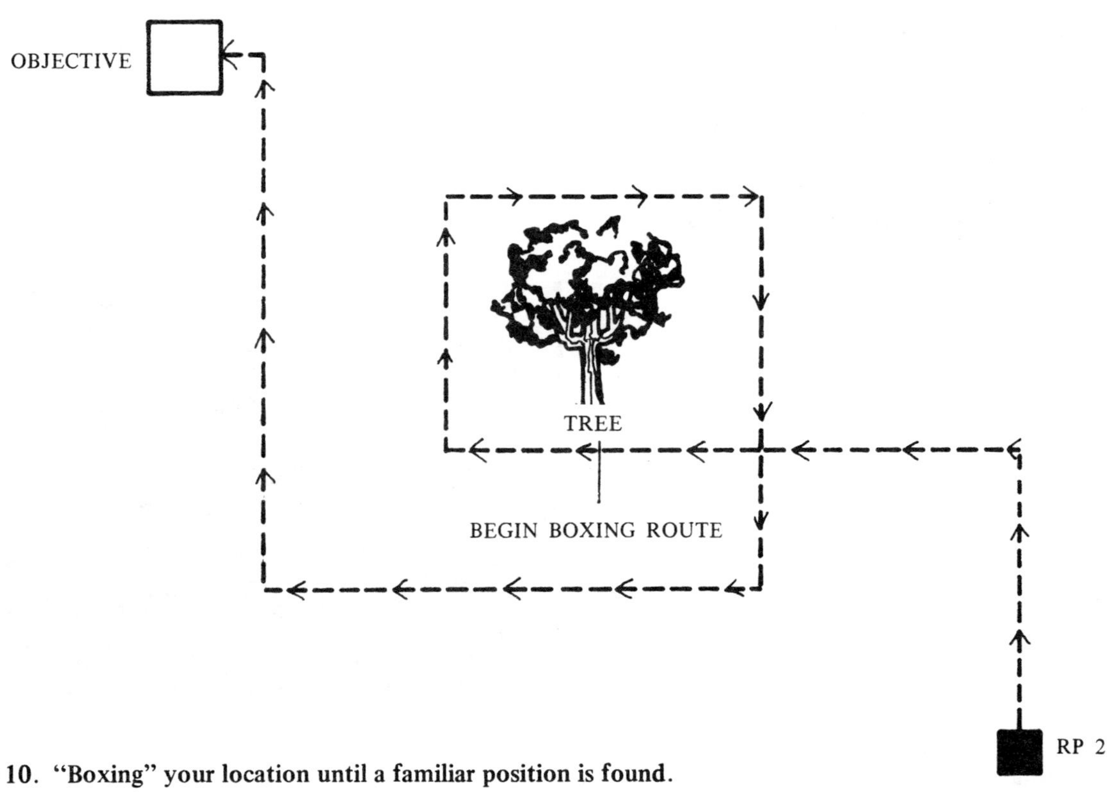

Figure 10. "Boxing" your location until a familiar position is found.

Remember that night navigation can be learned by anyone; it is simply a matter of practicing and applying simple skills in a logical order. Your subordinates and messengers may need lots of practice to master the skills, but every one of them can learn to navigate efficiently at night.

MAJOR DISORIENTATION

If you escape from a POW camp or if your aircraft is downed enroute, "boxing" your present location may not be practical; the distances you would have to travel are too great. You will be able to select a general direction and maintain it, but do not expect to glance at the stars and march confidently back to a U.S. position fifty kilometers away!

As you travel, even as a POW, try to remain aware of your general location and the direction of travel. Ask, "Where was the sun (or North Star) today? What direction did we travel? How far have we gone?" If you know the approximate location of where you started your travel or were captured and if you pay attention to the direction and approximate distance you subsequently travel, then you will have a good idea of your new location and the distance and direction you must travel to get back to friendly lines. Studying a small-scale map (which shows a large area) of your AO and the areas surrounding it when you first get to a new AO will help.

Figure 11 shows the North Star and the "marker" star groups, the Big Dipper and Cassiopeia. The North Star is in a constant position, but the markers rotate around it. The Southern Cross, in the Southern Hemisphere, is more difficult to use but provides a good south landmark. Locate the Southern Cross and extend the long axis five times its own length. The imaginary point at the end of the extension is south (see Figure 12).

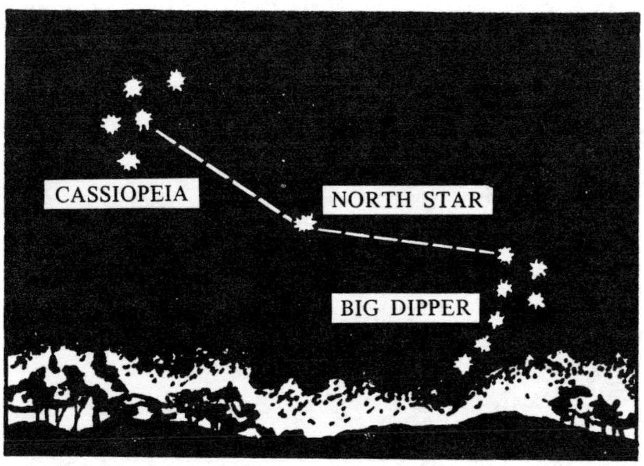

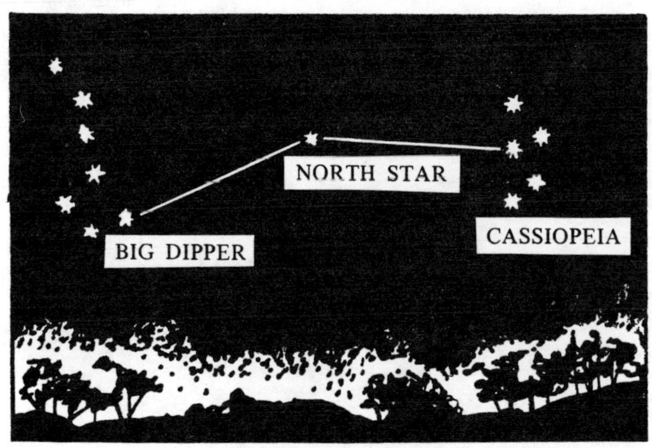

Figure 11. Locating the North Star for directional reference (in Northern Hemisphere). The North Star is in a constant position, but the markers rotate around it depending on the date and time.

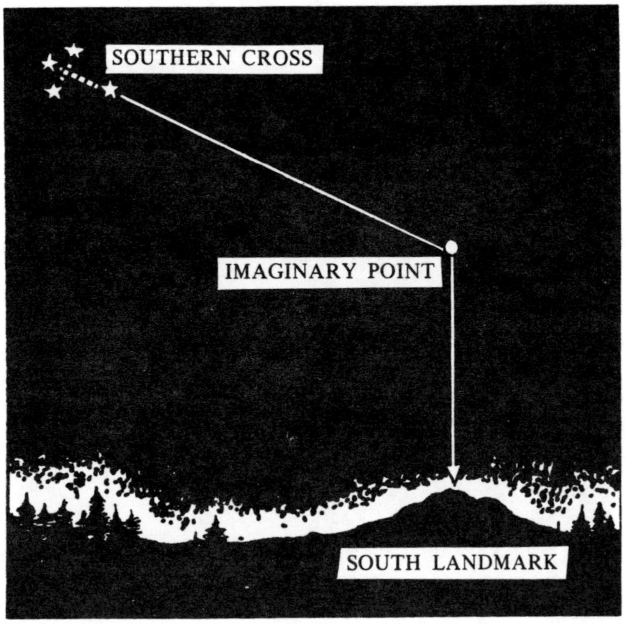

Figure 12. Using Southern Cross for directional reference (in the Southern Hemisphere).

Aviation survival kits often contain a small-scale map, printed on nylon, of the area. This map can sometimes be oriented to the terrain quite easily, and you can determine your location and the best route to travel.

"There seems to be a town west of us, with two highways running northeast from it and a railroad crossing through the town north to south. Where does something like that appear on the map?"

"The map shows a big river running southwest, through a canyon, from a mountain range into a broad valley. We're in a broad valley; is there a big river coming out of the mountains north of here?"

The bibliography lists several easy-to-read, highly informative books on this subject. Some of them can be found at the local library.

5. Target Detection and Engagement

TARGET DETECTION

One of the keys to target detection at night is to know your enemy. Think about his techniques and study his positions during recon patrols. Most unit positions form a pattern. When you understand the pattern, you know where to look for others after you find a single position.

One detection technique is to look for sentries/positions at the head of gulches, overlooking bridges and obstacles, and on the military crest of prominent terrain—emplaced for maximum observation. Then find positions that support the first one. Remember the ranges of supporting weapons and NVDs and search that distance from the known position very thoroughly, looking for indicators of enemy activity.

Think "patterns" and keep your mind blank; look calmly and methodically through the area; don't focus your eyes or attention on the surface, but on the patterns, noticing straight lines and any "strange" pattern. Then think "tone variations" and, keeping your mind blank, repeat the search. Continue through each of the appropriate indicators listed below. Patience, confidence, and calmness are critical for effective sensing, and they are all improved by practice.

When looking for indicators of enemy activities, you can borrow from the techniques of professional trackers who observe an area for:

Displacement. Are stones moved, crumbled (leaving lighter-colored faces and chips), or pressed into the ground, leaving a "rim" around them? Are bushes or grass bent? Watch out for ambush, especially if you are using artificial lights to look for these indicators. Send security well ahead and to the flanks as you search for enemy troops and positions, especially moving or patrolling troops.

Staining. Look for blood (which may look like tar or mud or even disappear in filtered light), water on stones, and crushed leaves. If IR is available, use it; practice with it in training. For example, broken and crushed vegetation gives a different signature than growing bushes; vehicle yards can be sensed at a distance; and—sometimes—previously occupied personnel positions can be detected. Freshly wetted rocks and newly dropped blood on bushes can be sensed with your fingers, as well as with IR devices.

Weathering. This indicator refers to the changes in signs due to the effects of weather and is not easy to detect without light and lots of experience. Its primary value is as a measure of the age of other indicators.

Littering is evidence of distracted or undisciplined troops. Watch for booby traps hidden in the litter.

Camouflage. Watch for straight lines that are rarely found in nature; contrasts in color or tone; unnatural vegetation such as green leaves among dead branches. Use IR to identify cut foliage. Get down low and look up to check outlines. Look along the ground, under vegetation, for uncamouflaged articles.

Sounds. Place your ear on the ground or on a stick driven six inches into the ground; it's hard to determine direction, but sounds can be heard from a long distance because the ground is denser than air and carries sound waves more efficiently. In light mist, sound will carry farther; however, sounds can be masked by dripping, rain, or wind. Also, rain may cause your targets to seek shelter and move less.

Rain may cause your men to miss sounds due to earflaps or rainhoods. Manage the misery; rotate listening duties during a stop. During stealthy movement, all men should be listening; that is, no helmets, rainhoods, or earflaps (helmets make hollow sounds in the rain, whistle during breezes, and generally rattle and rustle, masking important sounds).

If a dullness or ringing in the ears occurs, it may be due to imbalance between your outer and inner ear. Correct this by pinching your nostrils and blowing gently against the pressure. If the problem is increased, gently alternate inhaling and exhaling against the pressure until your ears "pop."

Odor can be an indicator in some situations. Sounds and odors are carried well by light breezes; with no other breeze, an effective air current too light to feel will move downhill in the evening, uphill in the morning. This is caused by the sun heating and cooling the air; air heated by the sun rises; as it cools when the sun goes down, it drains downhill into valleys.

Detect light breezes by wetting one finger on each hand. Determine which side dries first on both fingers—that direction is upwind. A sense of coolness—on the face, for example—can be misleading because it depends on the sensitivity and dampness of the skin in various areas.

Hint: Cup hands behind the ears and listen upwind. Hold your nose 45 degrees upwind and sniff. Sounds and odors may arrive intermittently.

See FM 90-5, *Jungle Operations,* for a more thorough discussion of tracking and indicators.

Distances

The distances at which the various indicators can be seen, smelled, or heard depend on an individual's natural ability and on the amount of water vapor in the air. Clear, dry desert air will allow you to detect sights at a greater distance, but does not carry sound or odor as well as more humid air. Since the amount of water vapor carried in the air can vary from hour to hour and place to place, you should check your environment continuously. Remember: If you can detect signs of the enemy at a particular distance, *he* may be able to detect signs of *you* at the same distance. Exercising care to remain undetected will help you stay alive and win the battle.

The following lists give rough estimates of detection distances at night.

Seeing. Varies with the amount of water vapor in the air and the amount of "background" light:

Headlights:	4 to 8 km.
Muzzle flashes—rifle:	1 to 5 km.
Muzzle flashes—cannon:	1 to 5 km.
Bonfire:	6 to 8 km.
Flashlight, unshielded:	1 to 2 km.
Lighted match:	100 m. to 1.5 km.
Lighted cigarette:	300 m. to 800 m.

Heat waves rise above a heated object, causing the stars to appear to shimmer. This can be an important aid to precisely locating vehicles at night, even if they aren't running.

Hearing. The distances over which sounds can be heard are extremely variable, but are generally greater over water and flat land and much less over or through vegetation. Also, sounds are deflected—they "bounce"—from buildings and rocky terrain and can mislead you about the direction of their source.

Rifle fire:	up to 2 km.
Normal speech:	up to 500 m.
Digging:	up to 1 km.
Engines:	up to 3 km.

Sound travels about 300 meters per second, so count the seconds between seeing the flash of a weapon firing and hearing its report. The number of seconds multiplied by three equals the distance to the weapon in hundreds of meters. For example, let's say you see the flash of a howitzer firing and count, "one one thousand, two one thousand, three one thousand," and then hear the report. That howitzer is approximately 900 meters away from you.

Smelling. Detection distances are variable and depend on an individual's odor sensitivity:

Diesel fuel:	up to 500 m.
Cigarette:	up to 150 m.

Odors from people with differing diets may help you identify friend or foe. For example, odors from meat-eaters, cheese-eaters, and vegetarians are fairly easy to tell apart. Other odors help identify enemy—or your—positions as well.

Practice builds confidence and skill—and skill can be developed with practice. Be sure to give your men feedback ("that vehicle you can smell is 100 meters upwind") so they can improve and build confidence.

Practice listening and sniffing while wearing blindfolds to build skill and confidence quickly.

TARGET ENGAGEMENT

Rifles

Firing with optical aids is similar to day firing. In general, use the least sensitive settings to avoid flare in electronic devices. You may want the assistant gunner to use the NVD and adjust the gunner's fire to avoid losing targets in the flare of muzzle flash.

Much of night firing must be done by feel. To convince your men that this method is effective, have them look at a chemlite on a tree, close their eyes, and point at it with the index finger of their strongest hand. When they open their dominant eye, they'll find their finger points at the chemlite. They'll find that pointing with their weaker hand is not quite as accurate because they use their weaker hand to point less often. They can learn to point with their rifle as accurately as they do with their finger. They can hit silhouette targets to about 50 meters consistently by pointing their weapon properly.

Human brains rely most on information from the dominant eye. To determine which eye is dominant, point at a small object, such as a doorknob or distant stump. Still pointing, close one eye, then the other. While looking through the dominant eye, the object appears to stay in alignment with your finger; while looking through the other, the object appears displaced. Every man in your squad should know which of his eyes is dominant, for he will practice point-shooting with his dominant eye directly over the rifle.

Pointing Practice

Don't waste the time you spend on a march by simply carrying your rifle—point with it and gesture with it. Have your men practice pointing at birds, windblown debris, and even thrown targets for a portion of every hour that they spend marching nontactically.

Soldiers tend to shoot high, often very high, when they fire at night. So, practice shooting low—very low!

Firing by Pointing

To fire, your men should shoulder their weapons and, with their dominant eye about two inches above and directly over their rifles, concentrate on the base of the target. If they look at the center of the target, they'll tend to fire high, and if they align the wrong eye—or neither eye—with the rifle, they'll fire to one side of the target.

They should fire as soon as the rifle is pointed at the target with a sharp pull on the trigger, not with a steady squeeze, which will disrupt their concentration and the rifle's alignment.

Teach your men to "look hard" at the base of their target (or where they think the base is, if they can't see it); if they try to aim with the tip of the barrel, they'll shoot high.

Shooting by pointing is much like shooting clay pigeons with a shotgun, so adopt the techniques the champion trapshooters use when they dry fire. Champion trapshooters spend up to an hour a day shouldering their weapon and following a moving target. Many of them practice while dancing to music to get used to concentrating despite distractions, and shooting (pointing) effectively often while off balance or facing away from their targets.

Reminder

Inexperienced soldiers fire quite high and enough to one side to miss, even with MILES. To train good night fighters, supervise their practice. Make sure each man can shoulder and immediately fire his weapon while looking hard at the base of his target.

You can check a squad member's technique by observing from one side to see if his rifle is either horizontal or the muzzle is slightly depressed when he fires. If it is not depressed, have the man stare at a spot, close his eyes, and point at the spot, first with his finger, then with his rifle. When he has developed a "feel" for pointing his rifle and increased confidence, start him back on the point-shooting exercise. Sometimes having him move his hand forward on the stock corrects the pointing-high error quickly.

Keep in mind that focusing the center of the eye on a target at night will cause the target to disappear because of the scarcity of night vision cells in the center of the eye (see Chapter 1 for more information); so "look hard" but don't stare. "Looking hard" is more concentration than simply staring. And, during practice, if your men visualize themselves succeeding—hitting their target—they'll do far better (see "Visualization" in Chapter 8).

Support Fire

Fewer than one-half of the men in an average infantry squad fire their weapons during any firefight. Fear and mental inertia are considered the most likely reasons for this. During the Civil War, over half of the rifles recovered had at least two charges loaded on top of the main round; the soldiers reloaded, but didn't fire!

Since light fighters are well above average, every member of your squad should fire in a firefight. But, your men may need a rule to follow to get them out of their shock and mental inertia. One rule that works for some elite allied units is: "When supporting by fire, count as if you were marching and fire every time your left foot should hit the ground. "Hup (fire), two, three (fire), four, hup (fire) . . ."

If you require your men to fire in this rhythm and ensure that they have developed the "pointing instinct," they will automatically shoot at enemies. They will think and shoot in a controlled, rhythmic manner. This means that the team providing cover fire will devastate a target area, freeing the maneuver team to do their job more safely. If all of your men fire, you'll have twice the firepower of an average unit!

Saturating with Fire

You may want to direct your squad's fire into a suspected area. In daylight, voice commands and hand signals are effective. However, at night, determining and communicating direction of fire is often difficult.

One aid is another rule: "In a firefight, search your sector for a target to shoot. If you don't find one, look in the next sector toward the shooting and shoot two rounds, but then search your sector again, immediately."

Another help is the use of a marker, such as a chemlite in each sector or every other sector. Then you can direct, "ten rounds to the left of the red sector" to saturate an area with fire.

You can also direct fire into an area by example: Order "Ten rounds with the tracers" and then fire tracers into the sector. Or, you may want to tape a chemlite to a rock and throw it in the direction you want your men to fire. Or, spill the fluid from chemlites on a handful of rocks and put each rock in a clear bag; direct fire by throwing the lighted package toward where you want your men to fire.

Hints: To conserve ammo, specify the number of rounds your men should fire. Also, ensure that your men immediately check their own sectors after firing into another.

Selecting Targets

Develop an SOP to help your squad establish priorities in selecting targets. For example, "AW and grenades engage AW; if no AW targets are within your view, fire at seen riflemen, but not at muzzle flashes alone"; or "Rifles engage seen targets only"; or "Rifles fire no more than two rounds at a muzzle flash—don't fire again until you see the flash again." The principle is to first engage the targets that hinder or threaten your unit most. Leaders must verbally establish and re-establish target priorities in a firefight; SOP instructions help keep your men firing but you must direct the firefight.

Marks to Locate Targets or Possible Targets

It may be difficult for a rifleman to remember exactly where he thought he saw a target, and harder yet to explain to you where he saw it. In daylight, he can use terrain features: "About 50 meters left of the black stump, the one on the left, by the pine tree." In the dark he can describe directions by using a range card: "From my position, look about three fist-widths right of my left-sector stake and about one fist above the sandbags."

You may want to put "spots," reflective or luminous marks, on trees or bushes in the sectors before dark. Usually a three-spot by two-spot grid in each sector is sufficient (see Figure 13). Then your riflemen can locate a possible target as "over two spots from the left and halfway down from the bottom one." Also, knowing the height of the spots, they'll be better able to fire low and effectively.

Remind your men to memorize the area in front of them so they can picture it at night. The luminous dots

Figure 13. Luminous dot grid for target-direction reference.

may help them keep track of where features are. Then, if they hear footsteps in mud or in dry bushes, for instance, they can picture the area and know where the mud or bushes lie relative to their position. Often you won't be able to see targets, but you'll be able to direct fire on likely enemy locations—if you memorize the area.

• Inspect your men's ammo frequently—corroded cartridges cause malfunctions.

• Watch your MG crews; many tend to "fiddle" nervously with ammo bandoleers, making a distinctive, compromising sound, as they walk or wait.

Hand Grenades

Hand grenades are often the weapon of choice at night. They don't give your location away and are effective even if they miss because they can still stun and surprise your opponents.

At night, be particularly careful about overhead obstacles; you don't want to bounce a grenade back into your position. Consider lobbing them underhand.

A sling adds leverage to hand grenades, increasing the range and, often, the accuracy. Fasten a cord about two-feet long to the grenade, release the spoon (let it "cook off" about one second), and then swing it toward your target. Be sure the cord is secure and don't waste time with several twirls; one is plenty to give the grenade maximum speed. This technique is a little tricky, but easy to learn, so practice it before you need it.

One method of fastening the cord is demonstrated in Figure 14. 1) Put the grenade in the toe of an old sock. Tie the cord around the body tightly, near the neck. 2) Split the sock to let the cord hang out while you fold the sock down over the tie, away from the neck. Tie the cord tightly around the grenade below the last tie—farther from the handle.

Another method is to tie the middle of the cord tightly around the neck. Pull the ends down and tie them in place with another cord near the base. Tape both ties tightly.

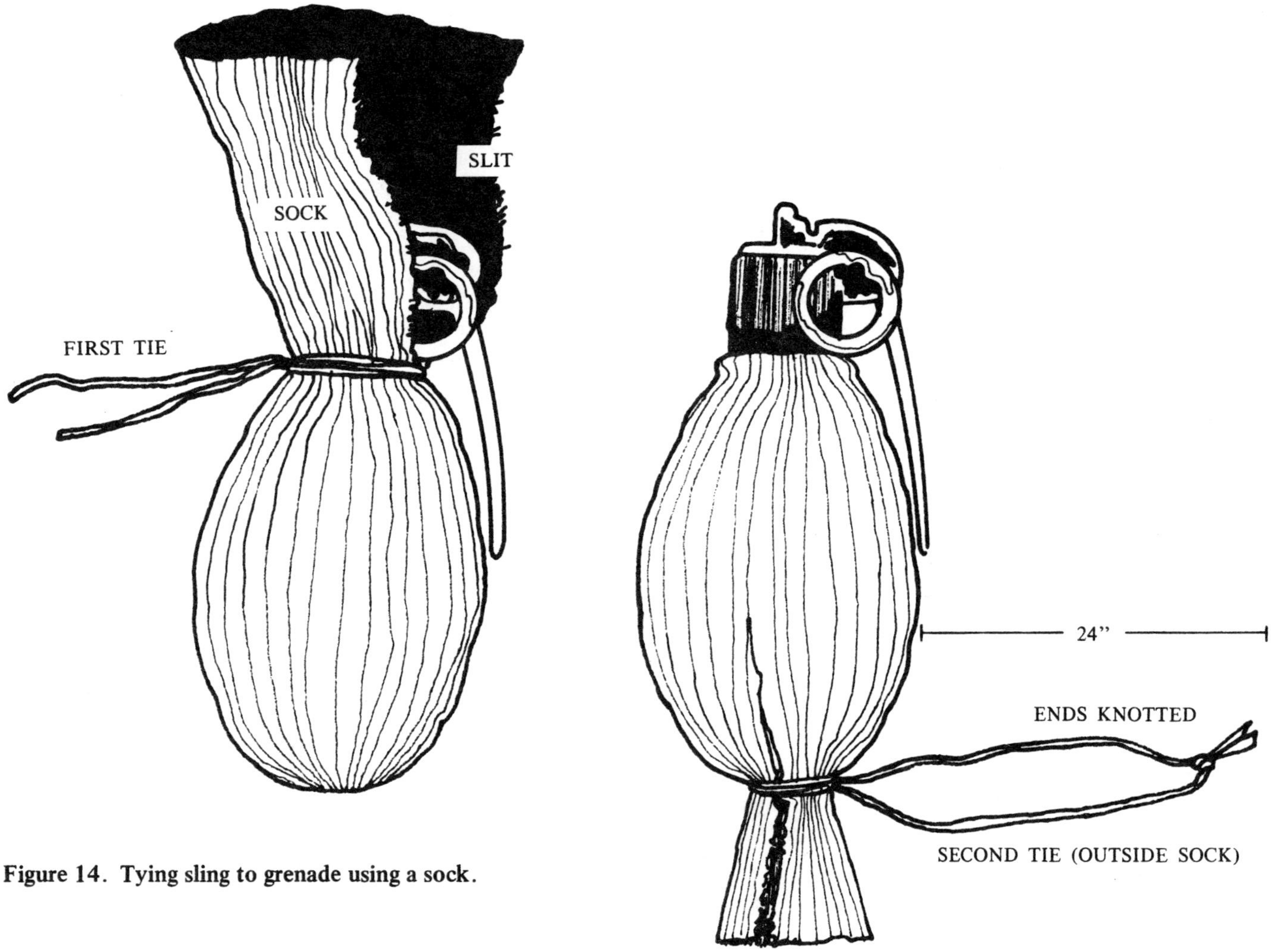

Figure 14. Tying sling to grenade using a sock.

Bayonets

Bayonets are especially effective in night fighting. They are silent, adding confusion and intimidation to your arsenal; they are effective in close-quarter fighting; and they provide a psychological advantage because bayonet fighters are universally feared.

The main physical problem with the use of bayonets at night is misjudging distance; the tendency is to thrust from too far away. Night practice on a bayonet course will help your men overcome this illusion of nearness.

The main psychological problem is timidity; new soldiers hesitate before attacking. Train light fighters to instantaneously lunge toward anyone who enters their position unannounced or toward any unidentified figure in the area. It is not necessary to get into a proper stance; that is, to check hand position, breathe, look hard, then lunge. That's too slow. Just lunge! If you don't have a bayonet or even a rifle, fine. You will probably figure out what to do when you get there and you will probably catch your opponent off guard, but if you hesitate, you will surely lose.

To ensure safety, you'll probably want to carefully enforce rules about wandering around the sector at night and identifying oneself before visiting another position. If both members of a buddy team hold the ends of a ten-foot rope, they can communicate. Each knows where his buddy is and they both can easily identify intruders. If they both fall asleep, though, the rope is a fine guide for infiltrators!

Bayonets as Fighting Knives

Place the bayonet across your hand diagonally, like a tennis racket. Grip it firmly with the thumb and first two fingers, loosely with the last three fingers (see Figure 15).

In hand-to-hand combat, use your free arm to deflect blows and keep your knife close to, and behind, your thigh in the correct grip. This protects the knife and adds intimidation to your attack by hiding the weapon (see Figure 16). Your enemy doesn't know what type of weapon you have or what you are going to do with it.

Figure 15. Hand grip for bayonet used as a fighting knife.

Figure 16. Balanced stance for using bayonet as a fighting knife at night.

Stab and slash only when close—within arm's reach. *Do not hesitate* (do something, even if it's wrong); try to close with and kill your enemy immediately, especially if he is larger or better armed. Aggression and speed are essential.

To silence a sentry with a fighting knife from the rear, tilt his chin up with your left hand, insert the knife into the front, lower part of his neck and then slash back and down—push in and pull back. If your victim is struggling and you cannot reach his neck immediately, insert your knife into the outside rear of his lower ribs and pull it toward his backbone.

The first method interrupts air flow below the larynx (voicebox) and severs major arteries; if thrust well into the chest cavity, the aorta will be severed, causing almost immediate unconsciousness.

The second method immobilizes and silences your victim through shock due to cutting his spleen and kidney and reduces the air pressure he can exert to yell. The victim, however, may struggle ineffectually for a minute or two; just hang on to him and continue pulling the knife toward his backbone.

One other point: Don't grab a sentry by his helmet; it may come off.

Expedient Silent Weapons

Silencer. A tube filled with metal washers, holes aligned, inside of a small engine muffler, is an excellent expedient silencer for a few rounds. Warning: Be sure the silencer is attached firmly and the washer holes are perfectly aligned or you may become the casualty; check the expedient silencer after every round. See Chapter 8.

Garrote. WD1 wire, one-meter long and firmly fastened to handles, can be used to choke a sentry to silence, unconsciousness, and eventual death (see Figure 17). If you are strong enough, you can use your arm power alone, otherwise loop the sentry's neck, turn back-to-back, and lever him off the ground by leaning forward—until he stops moving (see Figure 18). Warning: The tendency is to hurry, which can be a fatal error. Be certain the loop is around his neck, not his chin, and that he has stopped moving, not just stopped fighting for a minute. Also, see FM 21-150, *Combatives.*

Figure 17. Garrote made of wire with wooden handles.

Figure 18. Garrote attack.

Blackjack. A sock one-third full of sand (tied in place) can be used to stun a sentry. Hit him hard behind the ear. If he is needed for interrogation, use a club to the side of his neck, or else the middle of his back (between the shoulder blades). Be aware that he may not collapse, so follow up!

Bola. If silence isn't essential, an escaping enemy can be entangled with the cords of a bola. Bolas require considerable practice to become effective as weapons and are useful at night primarily in the open, such as on an urban street. See Chapter 8.

Warning: The bola will get you in more trouble than it's worth if you don't practice thoroughly with it before you need it. However, in the open, it is an extremely effective weapon to capture an escapee.

Indirect Fire at Night

Indirect fire (IF) adjustments at night are similar to day adjustments—you must specify changes in range and deviation. However, you won't see the flash or dust from the impacts, and the terrain features may not be visible, so sensing the round's impact is more difficult.

At night you can sense impacts by sound or sight. In either case: 1) remember the azimuth toward your target; 2) point at the impact of the round, state your sensing (short or over, left or right), and measure the azimuth of the direction you are pointing (remember to point—it's hard to remember exactly where you saw or heard an impact at night if you don't point when it occurs); 3) estimate the range correction, estimate or compute the deviation, and transmit your corrections.

Plan IF for the edges of minefields and obstacles like rivers; many inexperienced troops stop at these locations. Light fighters should stop only behind or ahead of these danger areas, crossing them quickly.

- Consider starting IF with illumination if your enemy freezes under illumination rounds.

- Consider making the last round of IF support fire WP (white phosphorus) so you'll be able to quickly assault behind the fire, knowing it has ended.

- If your enemy takes cover when IF fire arrives, consider preplanned, one-round fires into the enemy area at odd intervals.

- Notice—and be ready to report—enemy firing that isn't directed at you. You may be able to provide an azimuth to the weapon. This information can be used to locate and destroy it.

- See FM 90-5, *Jungle Operations,* for a practical discussion of adjusting indirect fire by sound.

CONCLUSION

Detecting and engaging targets at night depends mostly on your patience and attention to detail. So, as you train, practice thorough, methodical movements and target searches, and have your men practice until their rifles feel like extensions of their bodies. Your men will be as successful at night as you train them to be.

6. Illumination

Usually, night fighters fight without illumination. However, if illumination is available, you will want to ask yourself what to expect from it before you make specific plans. Adding an illumination plan as an afterthought or to "please the boss" can waste assets, and a poor plan may even hinder your operation.

Fully illuminated attacks are not simply daylight attacks at night, but they are similar. If your objective and route will be lighted throughout the operation, your control measures, signals, and personnel dispersion will be similar to daylight operations, but you can expect only as much light, at best, as you would have during full moonlight.

However, if the illumination is scheduled for only part of the operation, such as during your actions at the objective, you must plan to use night-fighting techniques during the other phases. You may move from your base, make a passage of lines, move to a release point, and begin the assault in darkness, then take your objective and consolidate under illumination.

Remember, though, that illumination is "iffy." It may be planned but whether it will really happen depends on *if* the weather is favorable, *if* the illumination support isn't diverted to another task–such as fire support–and *if* the illumination assets are sufficient to support your attack.

Thus, it pays to make an alternate plan for conducting your operation in darkness in case everything else doesn't work. You don't need to risk confusing your men by giving them both plans in detail; simply specify the general plan, "If we end up attacking without illumination, we'll . . ."

When you plan the use of illumination, first ask, "Do I want to recon–to determine the locations and patterns of roads, buildings, enemy positions, and natural terrain features? Do I want to simply observe the enemy troops' reaction to illumination? Do I want to locate their night (alternate) positions? Do I want to use the light to identify and engage targets and to aid in maintaining the direction of my attack? Do I want to mislead the enemy about my intentions, about what I really plan to do?"

With these questions answered, planning use of illumination is much easier. You can decide if you want light on call; for example, to reorient yourself by dropping a flare over a hilltop near your objective.

You can also plan locations to be lighted. Light should silhouette your targets instead of lighting them uniformly because uniform, or flat, light reduces contrast and makes targets hard to see. A flare over your route may even aid the enemy by silhouetting your men as they cross a ridge. Picture your route and objective and ask, "Where would I want a big flashlight turned on while I'm on this leg of the route, or on the objective?" Then plan to use illumination in those areas, lighting from the direction that gives you the most advantage.

Some considerations are:

- Overhead lighting, such as mortar or aircraft flares, is flat and of short duration but it illuminates defilades.

- Angled lighting, such as from a searchlight or a flare over an adjacent area, shadows defilades but emphasizes relief and makes target recognition easier. See Chapter 8 for ideas about demonstrating the effects of light.

Of course, light fighters often will have to rely on their own means and initiative to devise lighting. For example, mortars can fire illumination rounds, but this duty reduces their ability to provide fire support. Ask yourself, "What do I need more, light or fire support, during this operation?"

EXPEDIENT LIGHTING

- A salvaged headlight and batteries can provide effective neutralization of NVDs during attacks or in the defense. Remember that the enemy will probably destroy the light fairly quickly and can always move his NVDs. Also remember that the light may destroy your men's night vision. Usually, smoke is more effective at neutralizing NVDs.

- A thermite grenade in a can of liquid fuel will ignite a pile of brush or wood. Remember to ignite from the bottom of the pile and to use smaller wood lower in the pile so the flames will heat and ignite the large pieces. Don't use too much gasoline and don't pile wood directly over the (open) can of fuel for the initial flash could displace the pile (see Figure 19).

- Instead of wood you can use a covered drum of thickened fuel. Rig a detonating cord to cut the top off a half-buried drum and to fire a nearby—not inside the drum—WP grenade (see Figure 20).

- A Husch flare (see Figure 21) vaporizes fuel, then ignites the escaping vapor to produce a bright, fairly long-lasting light (about one hour). In this diagram, two mortar-round containers with 1/8-inch holes are wedged into a container of thickened fuel, but any similar devices can be used. The flare is ignited by a thermite grenade tied between the mortar containers (but not against them) at the level of the top of the fuel container.

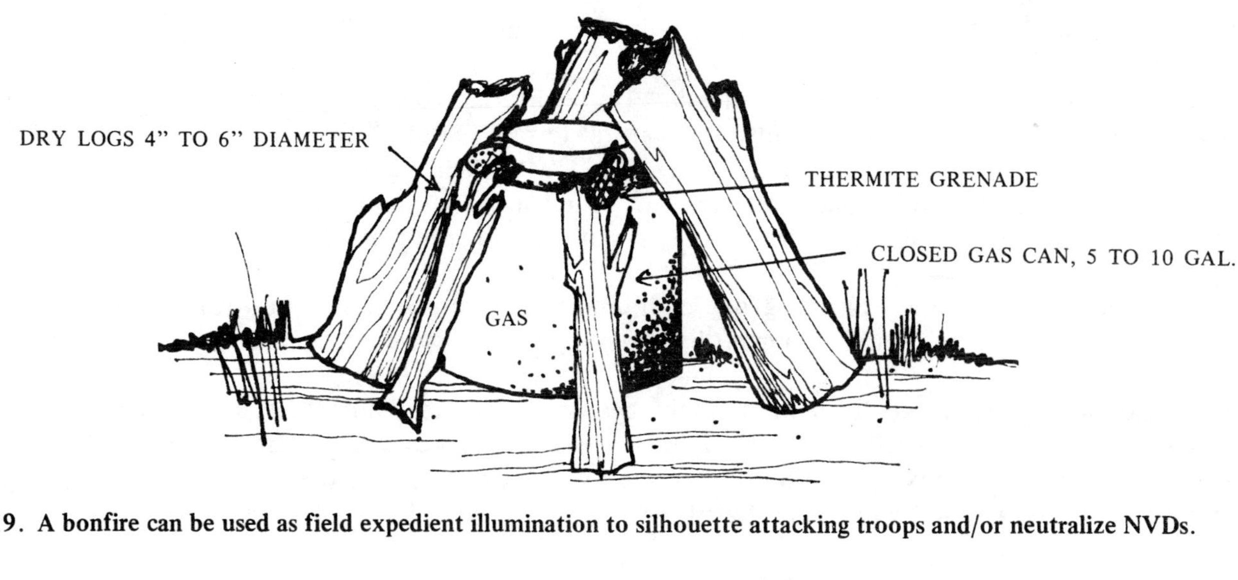

Figure 19. A bonfire can be used as field expedient illumination to silhouette attacking troops and/or neutralize NVDs.

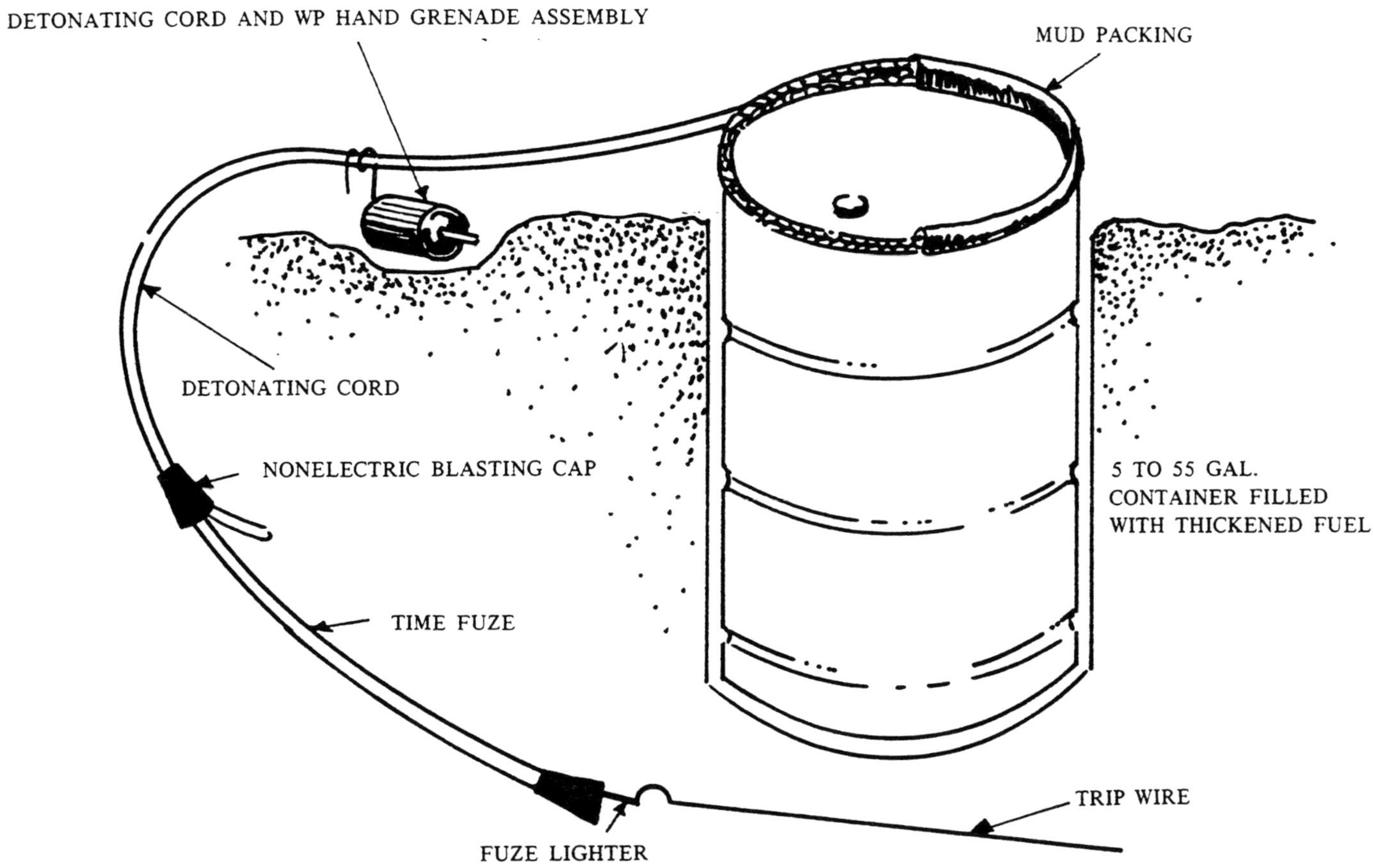

Figure 20. Expedient illumination using thickened fuel.

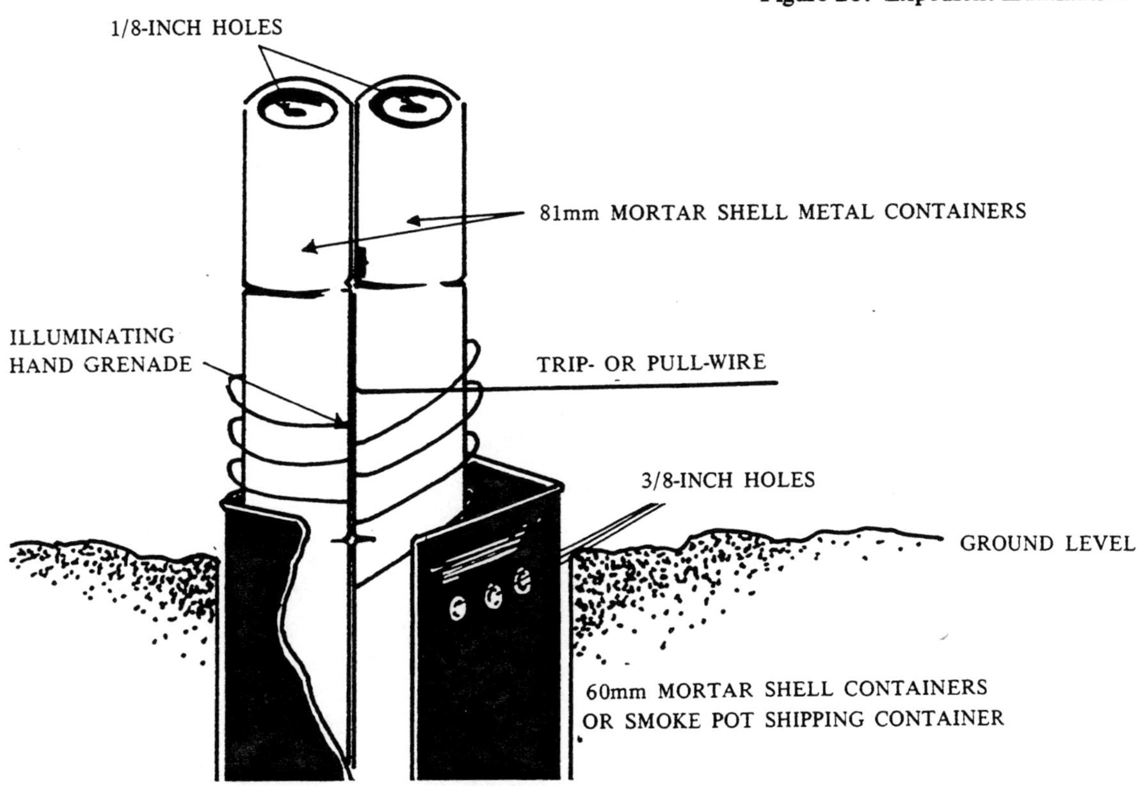

Figure 21. Using a Husch flare for expedient illumination.

Fuel, ignited by the thermite, heats the rest of the fuel, vaporizing it to stream through the mortar containers and out the 1/8-inch holes. Upon reaching the air, the fuel is ignited by flames from the fuel pot. The 3/8-inch holes in the fuel container allow the flames to be ventilated.

- Illumination with a 60mm mortar prior to attacking a tank is particularly effective. Light the tank from the front, if possible, to blind the crew, but not your tank killer team.

- WP rounds landing behind your enemy will silhouette him—and make a good target for you!

- Searchlights shined just above a man's height throw shadows that will identify your targets. This is especially useful in snowy terrain against arctic-camouflaged troops.

- Firing tracers into gasoline-soaked straw or brush can be used to "command detonate" illumination. Be sure, though, that you can find and hit the pile of straw.

- A can of fuel-soaked sand can light an area. Consider booby trapping and/or mining it since the fuel odor may give its position away. Be sure to report mines and booby traps in accordance with the SOPs.

- Try to plan illumination for the consolidation phase of an attack, in order to make searching for victims and positions easier and to help get your squad in position quickly to repel counterattacks.

- Rehearse the fire mission with the crews. Transition to and from illumination as well as support fire must be carefully practiced before the operation to avoid serious disruption of support. This is particularly true when you are using M203s and/or 60mm mortars for both missions. If possible, conduct the dry-fire rehearsal at the same time your squad rehearses so that communication problems can be identified before you begin the operation.

LIGHTING DEVICE EFFECTS

Type	Duration	Diameter Illuminated	Max Range	Advantages	Disadvantages
80mm mortar	75 sec	1,100m	3,300m	High-angle light	Must be refired for continuous illumination, flat lighting.
105mm howitzer	60 sec	1,000m	8,500m	" "	" "
155m howitzer	60–150 sec	2,000m	14,000m	" "	" "
Aircraft flare	150–200 sec	1,650m to 2,000m	N/A	" "	Requires aircraft to return to same point for subsequent flare drops, making it vulnerable, plus disadvantages listed above.
Searchlights					
IR	indefinite	varies with range	3,000m to 7,500m	Continuous, can be angular or from overhead (if reflected from clouds).	Strongly affected by weather, vulnerable to fire, attracts fire especially indirect mode.
Visible	"	"	3,000m to 1,500m		

7. Night Vision Devices

Night-seeing aids for light fighters fall into two general groups, techniques and devices. The techniques were described in Chapter 1, but we'll briefly review them here, since they are useful when using some types of NVDs.

First, to use the night-vision cells, which are concentrated away from the center of the retina (the seeing part of the eye), scan around what you want to see. Scanning throws the image of what you look at on fresh cells so the image won't fade when the cells get tired. Look to the side, top, or bottom of the object you want to see to focus the image on the parts of the eye that have the most night-vision cells. Most important, keep your mind calm. The lack of visual cues, and your imagination, can cause objects to appear to move, fade, and reappear—and can even cause hallucinations.

One of the most common night-vision aids are binoculars, which collect and concentrate light. Binoculars are mechanical light intensifiers. If you have a choice, select a pair with the lowest magnification and with the widest lens on the front. Navy binoculars are ideal, but the issue binoculars (7x50) have been found in tests to be nearly as effective as individual electronic image-intensifiers (PVS-295).

Whether using mechanical or electronic image-intensifiers, scanning and off-center vision can often improve your efficiency. Remember, though, that glass reflects light and could attract enemy attention, so hood your binoculars with a hat, brush, or your hand while observing. The hood is kept between the front lens and the strongest light source in front of your face, such as the moon or bright clouds. (Use an NVD such as a PVS-5 with binoculars to improve effective night-vision range considerably.)

Night-vision aids require maintenance and may need fresh batteries. Since they can be so useful in accomplishing your mission, their care should have priority—perhaps second only to your squad's weapons. Be sure your men protect them from shocks (such as dropping) and excessive heat.

Infrared NVDs sense invisible light energy—which is similar to heat energy—and form an image you can see. If you think of the devices as extremely sensitive heat sensors you'll be better able to use your IR devices and think of ways to neutralize your enemy's.

Most of the enemy forces that light fighters are likely to face are, in general, not as well equipped with NVDs. This doesn't mean that your specific opponents won't have many operating NVDs; it means that you might not be able to find as many, during a recon, as you would expect. However, it also implies that your enemy may have trained extensively without the devices; *don't underestimate him,* but don't attribute superior powers, either.

Higher-level units may have radar and sensing devices that you can use for a specific mission. For example, if ground surveillance radar (GSR) is available, you can coordinate with the operators to get periodic "fixes" of your location to help maintain control in the attack.

Remote sensing devices may be useful to you in a patrol base or while in a defensive position. The keys to using any device, though, are knowing what you want to accomplish with it and knowing how to employ, activate, and maintain it. Manuals for each device explain its use, and the operators or owners of the devices can brief you on the equipment.

Knowing what you want to accomplish starts with the questions you ask yourself. Start by visualizing the area you are going to attack or defend. Picture the landforms: "Where is an ambush likely? Where might I find enemy positions? Where will I have trouble maintaining direction? Where can I expect to start shooting? How will I know what to shoot at? Where do I most need to see?"

After you have identified the most critical areas to see, consider your NVD assets and match them up.

For example, if you want to see into a gulch that flanks your route and you have an aircraft with infrared or radar capabilities available, you only have to coordinate times, places, and communication. However, if you have only a PVS-2 (an NVD attached to a rifle), you'll have to ensure that the operator gets to, and searches, the gulch before you pass. Or, you might want your flank security to physically explore and secure the gulch.

In general, you'll want your longest-range NVDs observing toward your front or toward the most dangerous enemy avenue of approach. Next in priority, usually, is your flanks, with emphasis toward the most dangerous side. Generally, the leader should also use some form of NVD.

When moving, it's often useful to keep an NVD with the overwatch element (the unit watching the movement of another unit) and "leapfrog" it. So, you may have a device with your squad while on overwatch and then lose it later. Also, you may have to provide security for a higher-unit's NVDs in your area. Consider their defense as you would the defense of a crew-served weapon traveling with you. Their top priority may not be your squad's support, but they can be very helpful aids in your night operations. Helpful, that is, if you know what they can do and coordinate with their crews to get the information you need.

You may be able to use your shorter-range devices to aid in control of your unit. Here, too, planning is the key to success. Picture every step of your operation and plan to use your men, weapons, and NVDs to advantage on each step. Ask, "What will I want to do here? What will I do if I'm attacked from this direction? How will I get to that point in the dark?"

Be careful that the soldiers using NVDs don't "walk away" from your squad. They can see well and will tend to travel faster than the others if not controlled.

During the Falklands operation, a major factor in the successful attack by the British Third Parachute Battalion, particularly on Mount Longdon (11/12 June 1982), was neutralization of Argentine NVDs. The NVDs had been precisely located and were engaged by fire and artificial light at critical times.

Light fighters can enjoy similar successes by pinpointing enemy NVDs. Be sure you know where they are and, if possible, where they move when you shoot at them. The ranges at which NVDs can be disabled vary, depending on weather, terrain, and what is used to disable them. Even making them less useful helps, though. As a guide, consider the distances listed in Chapter 5 for unaided seeing and hearing at night as effective distances for neutralizing, but the closer the better.

Foil patterns and luminous markers are not particularly effective in disrupting a trained and alert NVD operator. However, remember that your opponents may well be tired and inexperienced with their devices. Also, even their temporary confusion can be to your advantage. Don't waste too much effort on trying to confuse operators with simulated "radio dials" (chemlite "juice" on a stick), positions (cigarette in a bush), or such; it's usually better to drop a WP round on or immediately in front of them than to try to mislead them. Use misleading indicators when no other options are immediately available, but remember that such tactics will expose your presence to enemy observation and fire.

A vehicle headlight and old battery can seriously disrupt image-intensifiers and have some effect on IR devices as well as destroy your enemy's night vision. Be sure it doesn't silhouette your men, though, or destroy their night vision. Also, ground flares, WP rounds, and sometimes air flares can be used to neutralize enemy devices if properly placed.

Fog or smoke reduces the range and effectiveness of image-intensifiers, but has only a slight effect on IR devices. However, both fog and smoke can increase your control problems, forcing you to close up your formation. If you plan to use smoke, or may have to pass through ground fog, plan ahead for simple, effective control measures, such as a buddy line between men, closer formations, and sound signals.

8. Training

Well-planned, properly executed training is the key to success in night operations. However, the small-unit leader cannot just give a lecture and expect his unit to subsequently perform properly. He must explain, demonstrate the proper way, rehearse during daylight, and practice, practice, practice at night. And, the worse the visibility, the better the training.

Before actually conducting training, the small-unit leader must *decide on a training program* (which is only another type of map showing how you get from one point to another and the checkpoints you use for reference), *evaluate the resources available* (time, equipment, facilities, manpower, etc.), *select those that best support the training,* and *provide for continuous evaluation of the training* (how well it works, what went wrong and why, how to do it better, etc.).

Much training for night operations can be integrated or concurrent with other training, both in the field and in the barracks. For example, your men can strengthen their night-walking muscles while they watch TV or read. Teach them to "sit" with their backs against a wall, buttocks about knee high. When they are correctly positioned, they will look as if they are sitting at the dinner table–without chairs. You can check their conditioning by having them "wall sit" during squad meetings and by holding impromptu contests in garrison.

The key is practice: practice during field problems, practice (blindfolded or in a dark room) in garrison, practice during breaks. Ask yourself, "What will we most need to know how to do in the dark, especially when we go into the field next time?" Your answer forms the task and condition, and implies the standards for your training.

Your training plan arranges available time, men, and material–available time, not "training schedule time." Men waiting for transportation can practice setting up a radio, cleaning a rifle, packing, or even decontaminating themselves–blindfolded. If you make a game of the practice, your men will be "having fun" instead of waiting and being bored.

Before you start a particular game, though, be sure you know what you want to accomplish. "Next break I want to practice rifle disassembly; each of my men should be able to disassemble and reassemble a rifle by feel so he can maintain it or troubleshoot problems in the dark. We can dismantle for 30 seconds, then hand the rifle and parts to the left and continue for one minute. We'll keep it up until the rifles are field-stripped, then assemble them by the numbers to learn to identify parts by their shape."

As another example, during a field exercise, have your unit move to a new location at 0200 hours. Do you let the unit "sack out" on top of a sleeping pad and "simulate" their fighting positions, "since we'll be moving in a couple of hours?" If you do, you're wasting an opportunity to practice getting ready and moving out silently—and without confusion.

In summary, situations and time for training abound, but the small-unit leader must know what he wants to train his unit to be able to do, if he is to take advantage of both scheduled time and small blocks of unexpected time. Further, he must be alert for opportunities to conduct integrated and concurrent training—training for several related, or sometimes even unrelated, tasks during one exercise.

PROBLEM SOLVING

When problems develop during practice, don't just solve them and drive on. Instead, analyze them. That is, ask questions until you understand all aspects of the problem. Once you have thoroughly described the problem in words, an answer is often obvious or at least easy to discover. For instance: A man sets off a flare while returning to his position after relieving himself. Quick solutions might include digging a fixed latrine, specifying an area for urinating, or telling the men to "hold it" or to "be more careful."

However, asking "What caused this problem? What is the worst thing that could happen in this situation? What are we trying to accomplish? How can we do it better?" leads you toward better solutions.

In this case: "The trooper was out in front, became disoriented coming back, and rushed, forgetting how to reorient himself or get help. If this had been combat, he could have been shot stumbling around outside of the perimeter, or hit by a mine if he missed the cleared path.

"If we stay inside the perimeter, we may develop sanitation and litter problems because of the crowding, but some of the men need more field time so that they don't panic if they're disoriented. If they leave the perimeter, they have to be able to return the same way, and they have to stay in touch with their buddy. Radios are out. Telephones are too much hassle.

"We'll try using a line. One buddy holds one end, the other takes the line out and ties it to his wrist while he's occupied. If he's in trouble, he can signal silently; he can follow it back, too. The biggest problem will probably be discipline, so I'll have to emphasize the rule in the beginning to make sure it's observed. If it works okay, we'll make it part of our SOP."

Some problems can be solved with a memory aid: "To tie a square knot, it's left over right, and around, then right over left and tighten." Be sure that each task is checked by feel, also; a soldier who can tie a square knot in the dark should be able to tell by feel whether he tied a square knot or a "granny."

SUGGESTIONS FOR FURTHER TRAINING

The remainder of this chapter presents a number of suggestions and tips for training the small unit in the techniques of night fighting. Remember to check the bibliography and other published guidance, such as field manuals, technical manuals, and training circulars. Also, use the men in your unit to help you train. For example, if one of your soldiers is especially interested in orienting, let him set up and run that training as a part of his professional development while you coach him on training techniques.

Training SOPs and Consistency

It is natural for the American soldier to express his unique individuality while in the army and, especially, in his dress and packing. However, the small-unit leader must develop commonsense SOPs and insist that his unit follow them. Perhaps the best way of accomplishing this task is to use the imagination and intelligence of your men to help develop the SOP in the first place. For example, you and your unit develop an

SOP to cover getting dressed and ready to move out at night in the field, then the unit rehearses it. The following is a sample walk through:

"We'll start from a sleeping bag, wearing underwear. Unzip the bag from inside, sit up, slip the BDU (battle dress uniform) jacket on, and button it. Grab the trousers, fly front, and pull your legs out of the bag and pull the trousers over your legs. Put on your socks, boots, hat, and webgear. Stand and fasten your trousers and webgear . . ."

When your unit has learned to dress and pack, time them—and announce the results. Test their habits when they are tired, under pressure, and when they least expect to be timed, such as in the billets. Test, time, inspect, and insist on improvement. Add interest by making the exercise a competitive game.

Other SOPs that are critically important include actions during consolidation of the objective, occupying a patrol base, security during halts, regular formation for moving forward, and actions at obstacles.

Also, you may want to specify priority for issue of equipment like NVDs; for example, first leaders, then gunners or assistant gunners of crew-served weapons, then point man, etc.

You might want to specify as part of the SOP the distance between men while moving, perhaps "within sight or sound of the man in front (or to the side) of you," and the distance for your point and flank security personnel to separate themselves from the main body (they should generally be within range of your control measures, of course).

Night Vision

The soldier's ability to see at night is critical to being able to perform well at night. Night vision (and all other senses) can be improved through training and practice. The various studies researching improvement through training disagree about the amount of improvement (50 to 200 percent), but all agree that ability does improve with training and practice. Some ways in which to train your unit might include the following.

In a large room with only very dim light seeping in from outside, have your men sit quietly and relax or meditate until their eyes have adjusted. Have arranged on the opposite end of the room various pictures and figures (small models of Soviet and allied equipment might work well). Have your men scan the room, using the appropriate techniques, and attempt to completely identify all of the pictures and objects. This will provide you with a baseline evaluation of your men's ability. Darken the room further and, if possible, change some or all of the objects. Repeat the exercise: relax, adapt to the darkness, identify the objects.

To demonstrate the effects of sudden flares of light on night vision, unexpectedly light a kitchen or wooden match holding it high enough for a few seconds so that it affects everyone (keep your own eyes closed). Extinguish the match and have everyone attempt to reidentify the objects. Readapt, have your men close their dominant eye, simulate the flare, and reidentify the objects. Discuss with the unit the time they needed for adaptation and readaptation, their individual abilities to adapt and recognize things, the effects of lights on adaptation, with and without eye(s) closed, the ability to discriminate colors, the effects of relaxation and meditation on night vision, etc.

The technique of carefully controlling an exercise provides the small-unit leader with a baseline evaluation of his soldier's individual abilities. In addition to helping you decide who needs how much of what type of training, it provides the answers to questions like "During tomorrow night's patrol, which of my men has the necessary night vision to be a point man? Who can I rotate at point?"

It will not, however, do much good for the unit or provide much training for an individual if you say,

"Smith, you have the worst night vision in the squad! I want you to walk point tonight to help improve it." This is negative training. By forcing this individual to stumble around in the dark, you reinforce his feelings of helplessness and inadequacy. If he does trip a flare or expose the unit, you add guilt feelings. Your training should always strive to reinforce or accentuate the positive. As a general rule, overcoming stress and adversity leads to a more confident, better trained individual. In training, it is the responsibility of the leader to manage that stress and adversity so that his soldiers succeed (positive training) and not fail (negative training).

Defensive Positions

Practicing digging-in can be combined with the stalking practice discussed in Chapter 2. Have one group try to locate and stalk the group digging-in, without detection. After a critique, change the group jobs; your soldiers will be able to avoid the errors the others made during the first practice. Be sure your men know that the men not digging are normally responsible for security; that is, they should be observing the area and ready to fire until their turn to dig.

Training in defensive operations is often a good time to conduct some of the night-sensing exercises in Chapter 1. You can combine night sensing with the digging-in practice by demonstrating sound travel at night while part of your unit practices digging-in. Men who are not digging can patrol outward from the position and report how far away they can hear the digging. These patrols develop your men's ability to distinguish the direction and distance of sounds.

When the patrol returns, ask the members questions that will help them fix what they experienced firmly in their minds: "Could you hear the digging in soft ground about 200 meters away from the position? How far did the sound of digging in hard ground travel? Did the sounds seem different? Which was harder? Softer? Could you hear voices? Could you hear as well this time as you could last time? What made the difference? Light mist makes the sounds carry farther, but muffles them; heavy rain covers sound, right? Did the brush and trees make any difference here? Did the sound of metal hitting rocks carry through the brush? How far away could you hear the digging when you put your ear on the ground?"

After the positions are completed, have the unit man them and then send out patrols to determine how far away a light can be seen, how far away smoking can be smelled, etc.

Each man must practice seeing, hearing, and smelling at night, over and over. Train them to make them aware of their senses and to improve their confidence and skill. It is important that every soldier in your unit practice and develop confidence in his ability to dig, fill sandbags, and build a position quietly in the dark; it's easier to "simulate" digging-in, but your men lose—a lot—if you hold them to loose standards while they train. "We'll fight like we train" is especially true about tedious tasks.

As your unit gains experience, you'll be able to develop SOPs the members can understand. If they know how far away conversation can be heard and cigarettes seen, they will more readily follow noise and light discipline procedures. In fact, a demonstration of how sound or light can compromise their safety may be an effective correction. For example:

"Jones, Smith, your talking could bring us attention we don't want. I want you to understand why. Recon north, toward that hill, past the streambed. Stop about every five minutes and listen, and when you return, report how far away you can hear our conversations. I will speak in a normal voice periodically. We'll listen to determine how quietly you move. After you cross the creekbed about 200 meters from here, carry on a normal conversation for two minutes; we'll listen for it. Then return quietly and report. We'll tell you how far away we could hear your approach."

While training in the field, you may want to have each man inspect another's site: both men will learn from the effort, and you can identify and correct mistakes as they inspect. For example, a well-meaning soldier may toss a rock he removed from his bed site into a bush to hide it. If the "inspector" misses the rock, point out the barren area, then have the soldier find and save the rock. Point out how far away the wet sides of rocks and the white ends of broken twigs can be seen.

Night Movement

In a classroom, you can practice night walking and assuming a prone position by using blindfolds. Simulate obstacles with chairs, brooms, and mops, and use bottle caps to help your men learn to step through and around loose rocks.

Combine night orientation with stalking by sending out a buddy team as a quarry. Have the quarry talk, smoke, or dig a position. If they spot a stalker, they flash him with a light to "kill" him. The objective of course is to stalk within arms reach of the quarry. This game can be played on a parade field in garrison as easily as in the field: Have the quarry team face away from the stalkers and turn around only when they hear a sound. They "kill" only the targets which they see moving when they turn around, so the stalkers will learn to freeze quickly when they are being observed.

Night Navigation

Accurate navigation, either by day or by night, is not especially difficult, but it is demanding. It demands that you accurately analyze the map, that you accurately plan your route, checkpoints, "catching features," etc., and that you pay close attention to bearings and distances during the march.

This section provides a brief review of Chapter 4. When planning training, the small-unit leader should take advantage of existing training facilities if they meet his training goals.

For example, a new squad leader wishes to begin training his squad in accurate night navigation; the squad does O.K. during day navigation exercises, but not well enough. The squad leader, SGT Smith, finds out that a large-scale relief model of Training Area 10 exists at the post. Using the model, SGT Smith explains to his squad the differences in day and night navigation; he has his squad squat down to view the model from the perspective of being on the actual ground. They use the model to plan their routes, checkpoints, catching features, etc. Using a penlight, SGT Smith moves his squad to Training Area 10 and navigates the planned course during the day, then at night. If possible, he actually uses his preplanned flares.

The key to training in this type of scenario is heavy involvement by the squad members in planning the course and in comparing the results of the actual ground movement with their plan.

1. Analyze terrain for relief (drainage). Ask: "What is the highest point in each direction? How much higher is it than my location? What are the highest and lowest points I can see in each direction from here? How far away is each point I can see from here? Which way would water run if it were poured on each point? Which way do the rivers run? Where are the most difficult areas to traverse? Where is the steepest and the flattest terrain in the area? Where are the roads in this area? Which way do they run? Do they appear on the map? Are there towns, farms, or airports nearby? Where? Which way does the wind blow? Does it change? When? Which way do houses face? Agricultural furrows and fences run generally in which directions? Do trees tend to lean in one direction? Where does the snow melt first? Which direction do the greenest hillsides face? What major feature can I see at night from here? From each high point in the area?" Not every question applies to every area, of course.

2. Route selection for night foot travel (priorities):

   a. *M*ission
   *E*nemy disposition
   *T*errain
   *T*roops
   *T*ime available

   b. Ease of navigation, silent movement, OCOKA, ease of travel, speed.

   c. For each leg (OCOKA):
   *O*bservation and fields of fire
   *C*over and concealment
   *O*bstacles
   *K*ey terrain
   *A*venues of approach

   d. To select a formation, consider:
   *N*earness of enemy
   *O*bstacles to travel and to your mission
   *S*ecurity while enroute
   *E*xposure to observation, fire

3. Navigation plan:

   a. Stay on a planned route by 1) navigating in a general direction using terrain features (terrain association), or 2) following a "compass" heading for a measured distance (dead reckoning). Remember to consider your destination while planning a heading.

   b. Checkpoints: Can I find them at night?

   c. Catching features: Linear, behind checkpoint.

   d. On a linear feature, use offset or bracket method with catching features.

4. Resection: Determine azimuths to two or more features you can identify on the map and on the ground; then plot back-azimuths on your map.

5. On the ground at night:

   a. On a finger or ridge, the ground slopes down in three directions and up in one (or is level in one).

   b. In a saddle between two hills, the ground slopes down in two directions and up in two directions.

   c. In a valley or gulch, the ground slopes up in three directions and down in one.

   d. On a hilltop, the ground slopes down in four directions.

6. Got lost plan:

   a. Preplan aids: lights (white or IR), MG tracers, indirect fire (WP, flares).

b. When lost, as soon as tactically possible:

1. Stop, get control of yourself and your men.
2. Think back: Where did I last know for sure where I was? What did I do next (direction, distance, turns)?
3. Predict: Where am I, probably?
4. Confirm: If I am in this location, what terrain features are near?

c. Resect from preplanned aids or prominent features.

d. Boxing: Travel a distance where you are still able to see the (marked) present location, turn 90 degrees by compass (do not guess), and travel the same distance again. Repeat, turn 90 degrees, and increase the distance you travel. Increase the side length every third side. Continue "boxing" around the point where you stopped until you find a location you can identify on both the map and the ground.

e. Confirm your location as if you were disoriented whenever calling for supporting fires or when you are asked by higher headquarters.

Target Engagement

You can check a squad member's technique by observing from one side to see if his rifle is horizontal or the muzzle is slightly depressed when he fires. If it's not level or slightly depressed, have the man stare at a spot, close his eyes, and point at the spot, first with his finger, then with his rifle. When he has developed a "feel" for pointing his rifle and has increased his confidence, let him try the point-shooting exercise again. Sometimes having him move his hand forward on the stock corrects the pointing-high error quickly.

Flashlight Pointing Exercises

Tape a small penlight to the muzzle of a rifle and run a wire from the switch to the trigger guard. Wire the flashlight so as to light when the wire touches the trigger guard. Make sure the light is taped to the top of the flash arrester and is aligned with the barrel.

Have your men "fire" first at stationary targets, such as a helmet about five meters away. Gradually increase the distance, and encourage them to fire sooner. When they light the target quickly, nearly every try, have them try to "shoot" a bouncing basketball, then a rolling baseball.

You can increase interest by challenging another squad to a shooting contest. To count, the target must be lighted when the light first flashes.

During live fire training, tape chemlites to targets for point-shooting practice. Fire one round at a time and if the target is definitely hit, count it as "one." If dirt is kicked up in front of the target, count it as "1/2"; if not, count the shot a miss, and consider three misses a fatal error. Tape over both front and rear sights to ensure that your men point shoot.

The use of spring-powered BB guns has been an effective aid in teaching point shooting and can be used in a suitable garrison location with reduced-size moving and stationary targets. With appropriate protection for your men (as a minimum, use goggles—BBs can cause severe eye injuries and blindness), you may be able to use existing buildings for MOUT training (techniques for urban fighting) or temporarily marked "lanes" in nearby training areas for outdoor point-shooting. (The new paint guns will probably not work as well for teaching rifle point-shooting; they are also expensive and messy.) *Warning: Do not attempt to use* CO_2 *or air-powered pellet guns—these can easily inflict serious or fatal wounds.*

Bolas

A bola is three weights, each on a line about two-feet long, with the lines attached. Rocks in socks make good weights, but handballs are better for practice bolas to reduce injury.

A bola is thrown by a right hander by holding the knot and swinging counter-clockwise until the weights are straight behind his head. He then throws the knot overhand, like a baseball, at a point one foot above and ahead of his quarry's knees. The bola is definitely not a silent weapon, but if skill is developed through practice, it can be handy for capturing prisoners in open areas, such as city streets.

Illumination

As an aid to visualizing the effects of the different types of lighting, the small-unit leader can use either an accurate large-scale terrain model or a very craggy rock. In darkness, use flashlights of various sizes and brightnesses to illuminate areas of the model or rock, varying the angle of illumination and the distance that the light is held from the rock to simulate the effects of illumination on terrain features.

Visualization

Visualization is a training technique used by many types of people, including great athletes and marksmen. It is a simple technique to learn and can be applied in many diverse situations. The keys to visualization are relaxing the body and focusing the mind.

Your mind is always active and usually cluttered with extraneous thoughts. By focusing the mind on one limited thought at a time, you reduce the "clutter" and bring all of your attention on that one thought.

To visualize, picture in your mind and feel in your body, one by one, all of the steps involved in a certain action. For example, competitive pistol shooters visualize a "good" shot. They *feel* a comfortable stance, the correct grip, correct sight picture, correct trigger pull, and correct let off and follow through; when they practice while standing, they can visualize so well that they can hold an imaginary pistol in exactly the same grip they would use with a real pistol, they move their trigger fingers exactly as they would in reality, and their bodies recoil exactly as they would.

Visualization can be applied to most facets of military life. For example, before a night patrol, visualize—picture and feel—starting off, the various legs of the patrol, the possible enemy actions and how you overcome them, the return to base, the debriefing, and the feeling of finishing a successful patrol. In the defense, visualize an enemy attack, "see" them charging your positions and your response, picture and feel direct and indirect fire on your position and your successful response, and visualize the battle ground after the attack and your feelings of a successful defense.

Have your men picture and feel their response under fire, "seeing" where an enemy soldier might hide and their "firing" at him.

By using visualization to picture and feel the various things that could happen and your successful responses to them, you are preparing your mind and body to cope with the shock of surprise. Sudden, unexpected enemy action provides a numbing shock to the mind and body; it is capable of paralyzing thought and leading to panic with disastrous consequences. Through visualization you can eliminate this surprise and shock, for you've already successfully coped with it in your mind.

HINTS FOR TRAINING

• Practice judging distances at night by throwing grenades at a flashlight. (An attached chemlite, fluid from a chemlite, or luminous tape makes practice grenades easier to find.) Use chemlites for live practice targets.

• Bayonet drills and stick fights should be practiced at night and on rough ground.

• Train in dim light to develop skills, then in complete (or nearly complete) darkness to master techniques.

• Consider using blindfolds to train in the light for night-fighting—you'll be able to supervise initial training more easily.

• Feedback (telling your men how well they performed) is critical. Whenever possible, use MILES or on-the-spot observers.

Bibliography

Antonov, A., Col. "Night Engagement." *Foreign Military Digest,* February 1974.

Betit, Eugene D., CPT. "Soviet Tactical Doctrine and Capabilities for Warfare at Night." Monograph. U.S. Army Institute for Advanced Russian and East European Studies.

Brown, Tom, Jr. *Tom Brown's Field Guide to City and Suburban Survival.* New York: Berkeley Books, 1984.

____. *Tom Brown's Field Guide to Wilderness Survival.* New York: Berkeley Books, 1983.

Buckhout, Paul, CPT. *Co C, 1/32d Inf(L) Tactical SOP,* 1984.

Buckwalter, Len. *The Pilot's Night Flying Handbook.* New York: Doubleday and Company, 1976.

Bornett, C., 1LT. *Training in Night Movements.* Translated from the Japanese. US Army Cavalry Association, Ft. Leavenworth, KS, 1914.

Conrad, H. M., LTC. "A Methodology for Wargaming STANO Systems." Master's thesis, Ft. Leavenworth, KS, 1971.

Coute, Kenneth R., Jr., 1LT. *Night Operations for Small Units.* Draft. Ft. Benning, GA, 1985.

Craft, Walter M., Jr., CPT. *POI for OSUT Personnel.* 1/32d Inf(L). Ft. Ord, CA, 1984.

Decker, CPT. "Night Operations: Lessons Learned." NTC briefing, 1983.

Drummand, R. J. M., LTC. "Night Attack, British Style." Paper on Falklands Islands Operations, available at Ft. Ord, CA, 1984.

Emanski, J. J., Jr. *Continuous Land Combat.* DLIC, Arlington, VA, 1977.

Farrell, James H. and Sternberg, Jack J. *Search Effectiveness with the Starlight Scope and 7x50 Binoculars.* Behavior Systems Research Library, 1970.

Infantry Journal Inc., *Infantry in Battle,* 1939.

Ivashentseu, N., COL, and Salinkov, Yu., LTC. "Specifics of Decontamination at Night," *Foreign Military Digest,* SOP, 1974.

LeMoyne, James M., LTC; VouDrie, Mark, CPT; and Sluder, Larry M., Jr., SFC. "MILES Game Equals Training," *Infantry* magazine, February 1985.

McKay, R.S. "The Shanghai School," *American Survival Guide,* February 1985.

Millican, R.W. and Hardman, W.F. *Restricted Visibility Land Combat Questionnaire Survey.* Advanced Research Projects Agency, Redstone Arsenal, AL, 1973.

Neal, Gilbert L. *A Provisional Care Curriculum for Infantry Night Operations Training.* Human Research Unit, Ft. Benning, GA, 1960.

O'Brien, James I., 1LT, and Stone, Clovis B., SFC. *Expert Infantryman Badge Digest.* Army Field Printing Plant, Ft. Benning, GA, 1964.

Phillip, George. *Signposts to the Stars.* London: George Phillip and Son, Ltd., 1941.

Phillips, William D., CPT. "The Light Leader's Course," *Infantry* magazine, January/February 1985.

Randolph, David. "Factors of Night Vision." Summary of lecture at Night Fighting Symposium, Ft. Hood, TX, 1983.

Sasso, Claude R., MAJ. "Soviet Night Operations in WWII." *Leavenworth Papers,* No. 6, December 1982.

Schneider, Robert J., MAJ, and Luscomb, Richard L., CPT. "Battle Stress Reaction and the US Army." *Military Medicine,* No. 149, February 1984.

Shparkovski, A., COL of MED SVC. "When Troops Operate at Night," *Foreign Military Digest.* February 1984.

Sport, William M., LTC. "Fact Sheet for Night Operations." ROK briefing, 1984.

Stein, Earl S. and Kobrick, John L. "Simulations, A Brief History." *Infantry* magazine. USAIS, Ft. Benning, GA, January/February 1985.

Usmanor, V., LTC. "[The] Skillfull Are Not Hindered at Night." *Foreign Military Digest,* September 1974.

Vats, V., Sr Lt. "Night Fire on Maneuvering Targets." *Foreign Military Digest,* September 1974.

Weitzman, Donald O. *A Survey of Some Human Factor Problems in Night Operations.* ARI, 1977.

MILITARY PUBLICATIONS (numeric order)

AFM 64-3; *Survival, Training Edition* (USAR). August 1979.

DA Pam 20-236; *Night Combat, An Historical Study,* June 1953.

DDI-1100-128-76; *Soviet Ground Forces Night Operations,* Defense Intelligence Agency, March 1976.

FC xxxx (draft); *Tactical Employment of Light Infantry,* USAIS (ATSH-B-ID), 1984.

FC 7-15; *Light Infantry Squad and Platoon Operations and ARTEP Mission Training Planning,* USAIS, December 1984.

FC 71-101; *Light Infantry Division Operations,* July 1984.

FC 90-1 (draft); "Night Operations." April 1984.

FM 1-204; *Night Flight.*

FM 20-60; *Battlefield Illumination,* January 1970.

FM 21-56; *Ranger Training and Ranger Operations,* January 1962.

FM 21-75; *Combat Training of the Individual Soldier and Patrolling,* 1957 and 1967.

FM 21-76; *Survival, Evasion and Escape,* March 1969.

FM 21-150; *Combatives,* December 1971.

FM 31-20; *Special Forces Techniques,* December 1965.

FM 31-36 (Test); *Night Operations,* April 1963.

FM 57-38; *Pathfinder Operations,* October 1968.

FM 71-2J (coordinating draft); "Tank and Infantry Battalion Task Force."

FM 90-5; *Jungle Operations,* August 1982.

FM 100-2-2 (coordinating draft); "Soviet Army Specialized Warfare and Rear Area Support," August 1982.

FM 100-5; *Operations,* 1976.

FM 101-5-1; *Operational Terms and Graphics,* March 1980.

FMFM 1-3A; *Field Firing Techniques* (USMC), 1980.

FMFM 1-3B; *Sniping* (USMC), 1981.

RB 90-11A; *Infantry Small Unit Cold Weather Combat Techniques,* January 1979.

ST 7-285; *STANO Handbook,* USAIS, Ft. Benning, GA, FY78.

ST 17-199 (draft); "Mounted Night Operations," US Armor School, Ft. Knox, KY, April 1983.

ST 21-75-3; *Dismounted Patrolling,* Ranger Dept., Ft. Benning, GA, January 1981.

TC 23-11; *Starlight Scope . . . Model 6060,* November 1966.

TC 23-13; *Crew-Served Weapon Night Vision Sight,* June 1967.

TC 23-14; *Sniper Training and Employment,* October 1969.

TC 23-18; *Night Observation Device, Medium Range (NODMR),* August 1967.

TC 350-1, Apdx G (2AD); *Night Operations Training,* March 1983.

8th IF G2; Comparison and Summary, Soviet Night Operations, 1973.

Human Performance in Continuous Operations, Vol. I and Vol. II, ARI, DTIC, December 1979.

The Marne Division, "Night Operations, A Great Promise Unfulfilled," Germany, 1982.

Military Intelligence, War Department, *German Military Training,* September 1942.

Night Fighting Symposium After Action Report, Ft. Hood, TX, August 1983.

Night Vision and Electro Optics Laboratory, *Presently Fielded Systems,* 1982.

Principles of Night Combat Parts 1, 2, 3, DTIC, 1955.

USAEUR Training Notes, *Highlighting Night Vision,* July 1932.